SpringerBriefs in Electrical and Computer Engineering

SpringerBriefs in Control, Automation and Robotics

SpringerBriefs in Control, Automation and Robotics presents concise summaries of theoretical research and practical applications. Featuring compact, authored volumes of 50 to 125 pages, the series covers a range of research, report and instructional content. Typical topics might include:

- a timely report of state-of-the art analytical techniques;
- a bridge between new research results published in journal articles and a contextual literature review;
- a novel development in control theory or state-of-the-art development in robotics;
- an in-depth case study or application example;
- a presentation of core concepts that students must understand in order to make independent contributions; or
- a summation/expansion of material presented at a recent workshop, symposium or keynote address.

SpringerBriefs in Control, Automation and Robotics allows authors to present their ideas and readers to absorb them with minimal time investment, and are published as part of Springer's e-Book collection, with millions of users worldwide. In addition, Briefs are available for individual print and electronic purchase.

Springer Briefs in a nutshell

- 50–125 published pages, including all tables, figures, and references;
- softcover binding;
- publication within 9–12 weeks after acceptance of complete manuscript;
- copyright is retained by author;
- authored titles only – no contributed titles; and
- versions in print, eBook, and MyCopy.

Indexed by Engineering Index.

Publishing Ethics: Researchers should conduct their research from research proposal to publication in line with best practices and codes of conduct of relevant professional bodies and/or national and international regulatory bodies. For more details on individual ethics matters please see: https://www.springer.com/gp/authors-editors/journal-author/journal-author-helpdesk/publishing-ethics/14214

Hardy L. C. P. Pinto • Liu Hsu •
Tiago Roux Oliveira

Sliding-Mode Control and Estimation for Time-Delay Systems

Fault Reconstruction, Disturbance Rejection, and Stabilization

Springer

Hardy L. C. P. Pinto
R&D Center (CENPES) - Safety Innovation Lab
Petroleo Brasileiro S.A.
Rio de Janeiro, Rio de Janeiro, Brazil

Liu Hsu
Department of Electrical Engineering
Federal University of Rio de Janeiro (COPPE/UFRJ)
Rio de Janeiro, Rio de Janeiro, Brazil

Tiago Roux Oliveira
Department of Electronics and Telecommunication Engineering
State University of Rio de Janeiro (UERJ)
Rio de Janeiro, Rio de Janeiro, Brazil

ISSN 2191-8112 ISSN 2191-8120 (electronic)
SpringerBriefs in Electrical and Computer Engineering
ISSN 2192-6786 ISSN 2192-6794 (electronic)
SpringerBriefs in Control, Automation and Robotics
ISBN 978-3-032-30177-2 ISBN 978-3-032-30178-9 (eBook)
https://doi.org/10.1007/978-3-032-30178-9

Mathematics Subject Classification: 34, 35, 39, 93

This Springer imprint is published by the registered company Springer Nature Switzerland AG
The registered company address is: Gewerbestrasse 11, 6330 Cham, Switzerland

Preface

Sliding Mode Control and Estimation for Time-Delay Systems: State and Input-Output Delays Perspectives. Time delays present one of the most significant challenges in modern control system design. In networked and distributed control systems, delays are not merely theoretical concerns but unavoidable realities that must be addressed to ensure stability and performance. Among these delays, *input and output delays* are particularly relevant—input delays occur between controller output and actuator implementation, while output delays arise between sensor measurements and their availability to the controller.

Sliding Mode Control (SMC), known for its robustness, offers a natural framework for handling such delayed systems. The key advantage of SMC—its complete insensitivity to matched uncertainties once the system reaches the sliding surface—makes it particularly attractive for time-delay systems. However, delays introduce significant complexity, as they can fundamentally alter the nature of the sliding motion, potentially undermining the robustness that makes SMC appealing.

This brief focuses specifically on the challenges and solutions related to *output delays* in sliding mode-controlled systems, although input delays could be analogously incorporated in some cases. While many texts have broadly addressed SMC and others have explored time-delay systems—mostly including those considering delayed states—there was no book in the literature dedicated to this specific topic. This volume provides a focused treatment, emphasizing these practically important delay types in different contexts, such as fault reconstruction, disturbance rejection, and stabilization.

Motivation and Scope Several key observations motivate this focused approach:

1. *Prevalence of Input or Output Delays:* These delays are common in real-world control implementations. Input delays arise from computational lags, communication networks, actuator dynamics, and signal processing, while output delays stem from sensor dynamics, data acquisition systems, and communication networks. These delays cannot be eliminated through engineering alone—they are intrinsic to the physical systems being controlled.

2. *Unique Challenges in SMC Design:* Input delays directly affect the control action, potentially introducing oscillations or even instability during the reaching phase, disrupting the sliding mode itself. Output delays, meanwhile, complicate state estimation and can lead to degraded performance even when stability is maintained. The central difficulty lies in realizing *(almost) ideal sliding modes*, which are crucial for ensuring SMC robustness in the presence of delays. This idea underlies the methods presented in this brief. In fact, one of the biggest obstacles in sliding mode control is *chattering* [20]—high-frequency vibrations that deteriorate system performance and can cause physical damage due to mechanical wear or excessive heat [21].
3. *Strategies to Overcome Chattering in Delayed Systems:* The presence of input or output delays requires special techniques to mitigate chattering [2, 4, 18]. Recent advances in predictor-based methods and robust observer design offer new avenues for addressing these specific classes of delays.

In particular, the evolution of predictor-based control for delayed systems is intertwined with early efforts to combine delay compensation and sliding-mode control, beginning with the predictor proposed in [16] and the subsequent clarifications in [7] and [3]. This exchange clarified a subtle but crucial point: once a predictor is introduced, disturbances that were originally matched with the control input effectively become unmatched in the transformed dynamics, thereby altering the classical robustness properties of sliding-mode control. This realization motivated later efforts on chattering reduction, disturbance attenuation, and more careful robustness analysis in predictor-based designs, notably synthesized in [14]. The perspective adopted in this brief builds on this historical understanding to revisit predictor-based sliding-mode strategies with an emphasis on disturbance rejection, fault reconstruction, and stabilization in time-delay systems.

These innovations deserve a coherent presentation that highlights their interconnections and practical applications. This brief covers design methodologies, stability analysis techniques, and practical implementation strategies. We progress from fundamental concepts to advanced topics, maintaining a clear focus on the specific challenges posed in particular by output delays. Depth is prioritized over breadth—we deliberately exclude discussions on distributed or classes of nonconstant delays, whose dependence is not related exclusively with respect to time (but also depending on the own state), unless they provide direct insight into output delay compensation.

Brief Organization The brief comprises three parts. **Part I** (Chaps. 1 and 2) deals with fault reconstruction via time-shifted sliding mode observers, **Part II** (Chaps. 3 and 4) with disturbance rejection via predictor feedback, and **Part III** (Chap. 5) with stabilization for time-delay systems by means of norm-state and cascade high-gain observers. These chapters are based on material originally published in [8, 10, 11, 12, 13].

Intended Audience This brief is primarily intended for graduate students, researchers, and practicing engineers working on robust and nonlinear control of time-delay systems. Readers are expected to have a basic understanding of control theory, state-space models, and Lyapunov stability. Some background in conventional sliding mode control is helpful but not essential, as we provide a literature overview in the corresponding sections "Notes and References" for each chapter and bibliography of fundamental concepts. We expect the following:

- *Students* will find a structured entry to an important subfield of robust control.
- *Researchers* will gain a comprehensive reference on the state of the art and key open problems.
- *Practitioners* will find actionable design methodologies and implementation guidelines tailored to real-world challenges.

Although the brief's material is of interest to applied mathematicians and engineers, the exposition is optimized for readers already familiar with basic notions of sliding mode, delay compensation, and solutions of discontinuous systems.

Notation and Examples Throughout this brief, we maintain consistent notation and provide examples that illustrate key concepts without excessive complexity. Whenever possible, we return to the same example systems across different chapters, allowing readers to compare alternative approaches to the same problem. The selected examples specifically highlight the challenges of *output delays* rather than general complexities.

In particular, the following notation and basic concepts are considered in this brief: **(a)** The maximum and minimum eigenvalues of a symmetric matrix P are denoted by $\lambda_{\max}(P)$, $\lambda_{\min}(P)$, whereas $\lambda_i(P)$ represents its i-th eigenvalue. **(b)** The Euclidean norm of a vector x and the corresponding induced norm of a matrix A are denoted by $\|x\|$ and $\|A\|$, respectively. **(c)** The symbol $\mathbb{E}[.]$ represents the expected value of a stochastic signal. **(d)** Classes $\mathcal{K}_\infty$ and $\mathcal{KL}$ functions are defined as in [6, p. 144]. **(e)** OSS means output-to-state-stable as in [19]. **(f)** In some parts of the brief, we invoke Filippov's definition for solution of discontinuous differential equations [1] due to the existence of the sliding modes and differential equations with discontinuous right-hand sides. In other specific cases, we need to invoke *generalized* Filippov solutions [17] due to the existence of the sliding modes as well as the mathematical technicalities involving a right-hand side exhibiting singularity at the prescribed convergence time instant. Although the applicability of Filippov solutions in infinite-dimensional systems governed by partial differential equations is not immediate, when needed we adopt an interpretation based on renormalization/regularization, as recently discussed in [15], or assume its existence employing the equivalent control method, such as in [7]. **(g)** As is usual in the time-delay literature [5, Sec. 1.2], the initial conditions of a system with state $z \in \mathbb{R}^n$ are given by $z(t) = z_0(t), t \in [-d_{\max}, 0]$, where $z_0(t)$ is a vector function continuous in $t \in [-d_{\max}, 0]$, and $d_{\max}$ is the maximum time delay. **(h)** A vector signal $\pi(t)$ is an exponentially decaying term dependent on the initial conditions $z_0(t)$, if $\exists k, \lambda > 0$ such that $\|\pi(t)\| \leq ke^{-\lambda t}z_0^*$, $\forall t \geq 0$, where $z_0^* = \sup\{\|z_0(t)\| : t \in [-d_{\max}, 0]\}$. **(i)** The symbol "$*$" denotes the convolution operator. **(j)** Assumption labels such as **(A1)**, **(A2)**, etc., are conveniently reused in each chapter and must be interpreted locally within the context of that chapter.

All theoretical developments are complemented by numerical simulations and, eventually, supported by *applications*. We believe that bridging theory and practice is crucial, particularly in a field where idealized mathematical solutions often encounter real-world implementation challenges.

Acknowledgments In the course of this research, we have benefited from interaction with Miroslav Krstić, José Paulo V. S. Cunha, Paulo Cesar Souza Silva, Victor Hugo Pereira Rodrigues, and Andrei G. H. Battistel. We would like to thank for their generous collaboration and valuable contribution in several parts of this brief.

We gratefully acknowledge the support that we have received from the Brazilian funding agencies CNPq, CAPES, and FAPERJ.

The authors would like to sincerely thank the Springer editorial and production team, especially Oliver Jackson, for their professionalism throughout the publication process. We also gratefully acknowledge Prof. Tamer Başşar for his continuous encouragement and support during the development of this brief.

Final Thoughts The field of *sliding mode control and estimation for time-delay systems* will continue to evolve rapidly. As mentioned before, this is the first brief or book in the literature dedicated to this specific topic, providing a snapshot of current understanding while laying foundations that will remain relevant as the field advances. Our goal is to balance rigor and accessibility, breadth and depth, and theoretical development and practical application. We hope this volume serves not only as a reference but also as inspiration for new research directions and practical implementations. As networked and distributed systems become more prevalent, the challenges posed by input and output delays will only grow. By providing specialized tools to tackle these issues, we aim to contribute to the broader field of robust nonlinear control in an increasingly complex and interconnected world.

Dedication Hardy's dedication is addressed to his parents and brothers, and, most especially, to his wife, Flávia, and his children, Eduardo and Juliana, for their love, support, and understanding.

Liu offers this brief in tribute to Professor Luiz Bevilacqua, from COPPE-UFRJ, with enduring gratitude for his vision, his generosity, and the open door that shaped a career.

Tiago dedicates this brief to the memory of his father, *Delci Teixeira de Oliveira* (1946–2024), and his mother *Vera Lúcia Carvalho Roux de Oliveira* (1945–2026), in gratitude for their life lessons and everlasting love. He extends this dedication to his wife, Deborah Roux, and his sister, Mirela Roux, for their constant affection and companionship.

Rio de Janeiro, Brazil
2026

Hardy L. C. P. Pinto
Liu Hsu
Tiago Roux Oliveira

References

1. Filippov AF (1964) Differential equations with discontinuous right-hand side. Amer Math Soc Transl 42(2):199–231
2. Fridman E, Fridman L, Shustin E (2000) Steady modes in relay control systems with time delay and periodic disturbances. ASME J Dyn Syst Measure Control 122(4):732–737
3. Fridman L, Acosta P, Polyakov A (2001) Robust eigenvalue assignment for uncertain delay control systems. IFAC Proc Vol 34:213–218
4. Fridman L, Fridman E, Shustin E (2002) Steady modes in relay systems with delay. In: Perruquetti W, Barbot JP (eds) Sliding mode control in engineering. Marcel Dekker, New York, pp 263–293
5. Gu K, Kharitonov VL, Chen J (2003) Stability of time-delay systems, 1st edn. Birkhäuser, Boston
6. Khalil HK (2002) Nonlinear systems, 3rd edn. Prentice Hall, Hoboken
7. Nguang SK (2001) Comments on "Robust stabilization of uncertain input-delay systems by sliding mode control with delay compensation". Automatica 37:1677
8. Oliveira TR, Cunha JPVS, Battistel A (2016) Global stability and simultaneous compensation of state and output delays for nonlinear systems via output-feedback sliding mode control. J Control Autom Electr Syst 27:608–620
9. Orlov Y (2008) Discontinuous systems: Lyapunov analysis and robust synthesis under uncertainty conditions, 1st edn. Springer, Berlin
10. Pinto HLCP, Oliveira TR, Hsu L (2019) Sliding mode observer for fault reconstruction of time-delay and sampled-output systems — a time shift approach. Automatica 106:390–400
11. Pinto HLCP, Oliveira TR, Hsu L (2019) Fault reconstruction for delay systems via least squares and time-shifted sliding mode observers. Asian J Control 21:52–69
12. Pinto HLCP, Oliveira TR, Hsu L, Krstic M (2024) Sliding mode control for disturbance rejection and estimation under measurement delay with PDE-backstepping predictor. Int J Robust Nonlinear Control 34:2865–2886
13. Pinto HLCP, Oliveira TR, Hsu L, Krstic M (2021) Sliding mode control for stabilization and disturbance rejection of uncertain systems with output delays via predictor and super-twisting observer. In: Proceedings of the 60th Conference on Decision and Control, Austin, TX, pp 3312–3317
14. Polyakov A (2012) Minimization of disturbances effects in time delay predictor-based sliding-mode control systems. J Franklin Inst 349:1380–1396
15. Polyakov A (2025) Finite-dimensional Filippov method for infinite-dimensional SMC system. In: Proceeding of the 64th Conference on Decision and Control, Rio de Janeiro, pp 6019–6024
16. Roh Y-H, Oh J-H (1999) Robust stabilization of uncertain input-delay systems by sliding mode control with delay compensation. Automatica 35:1861–1865

17. Seeber R (2023) Generalized Filippov solutions for systems with prescribed-time convergence. CoRR abs/2303.03133, https://doi.org/10.48550/arXiv.2303.03133, pp. 1–7
18. Shustin E, Fridman L, Fridman E, Castaos F (2008) Robust semiglobal stabilization of the second order system by relay feedback with an uncertain variable time delay. SIAM J Control Optim 47(1):196–217
19. Sontag ED, Wang Y (1997) Output-to-state stability and detectability of nonlinear systems. Syst Contr Lett 29(5):279–290
20. Utkin VI (1992) Sliding modes in control and optimization. Springer, Berlin
21. Utkin V, Guldner J, Shi J (1999) Sliding mode control in electromechanical systems. Taylor & Francis Ltd., London

Contents

Part III Stabilization

Part I
Fault Reconstruction

Chapter 1
Fault Reconstruction of Time-Delay and Sampled-Output Systems—a Time-Shift Approach

This chapter introduces a time-shift approach for reconstructing actuator faults in systems with output time delays using a sliding-mode observer (SMO). The methodology accommodates arbitrary output delay durations, even when uncertain and time varying, while ensuring ideal sliding-mode operation—critical for mitigating chattering effects and enhancing applicability in sampled-output systems. To facilitate smooth fault reconstruction via noncausal filters, an intentional output delay of one or more sampling periods is introduced.

1.1 Sliding Mode Observers for Fault Reconstruction with Measurement Delay

In this section, the time-shift approach is described for systems with measurement delay, which can be assumed arbitrarily long, time varying, and uncertain. Well-known results for observer design are then applied in a straightforward way to the time-shifted system under mild assumptions.

1.1.1 Problem Statement

Consider the following linear system presenting a time-varying measurement delay $\tau(t) \geq 0$:

$$\begin{cases} \dot{x}(t) = Ax(t) + Bu(t) + Ff(t) \\ y(t) = Cx(t - \tau(t)), \end{cases} \tag{1.1}$$

H. L. C. P. Pinto et al., *Sliding-Mode Control and Estimation for Time-Delay Systems*, SpringerBriefs in Control, Automation and Robotics,
https://doi.org/10.1007/978-3-032-30178-9_1

where $A \in \mathbb{R}^{n\times n}$, $B \in \mathbb{R}^{n\times m}$, $C \in \mathbb{R}^{p\times n}$, $F \in \mathbb{R}^{n\times q}$. The input signal $u(t) \in \mathbb{R}^m$ is known, the system state $x(t) \in \mathbb{R}^n$ is not measurable, and $y(t) \in \mathbb{R}^p$ is the output signal. The unknown disturbance signal $f(t) \in \mathbb{R}^q$, $q \le p < n$, represents the effect of an actuator fault.

The main objective of this chapter is to develop a method to reconstruct the actuator fault signal $f(t)$ based only on the known input and output signal $y(t)$, which is measured with delay $\tau(t)$.

The following assumptions are made:

(A1) Matrices A, B, C, F are known. B and F have full rank.

(A2) The time delay $\tau(t)$ is differentiable and satisfies $0 < \tau(t) \le \bar{\tau}$ for some constant $\bar{\tau}$, and its time derivative $\dot{\tau}(t)$ satisfies

$$\underline{d} \le \dot{\tau}(t) \le d < 1, \quad |\ddot{\tau}(t)| \le r_d \tag{1.2}$$

where r_d, d, and $\underline{d}$ are constants, with $\underline{d}$ being possibly negative and d sufficiently small. These conditions must be simultaneously satisfied.

(A3) The unknown disturbance/fault $f(t)$ is bounded by

$$\|f(t)\| \le \alpha, \quad \alpha \in \mathbb{R}^+. \tag{1.3}$$

(A4) The pair (A,C) is observable, and the invariant zeros of the triple (A, F, C) lie in the left half plane.

(A5) $rank(CF) = rank(F) = q$.

(A6) The past values of $u(t)$ for $t \in [t-\tau, t]$ are available for the fault reconstruction method.

1.1.2 Time-Shift Approach for Observer Design

From **(A4)** and **(A5)**, it is assumed that the system (1.1) is already in the following canonical form suitable for observer design [11, Lemma 6.1]:

$$A = \left[\begin{array}{c|c} A_{11} & A_{12} \\ \hline A_{21} & A_{22} \end{array}\right], \quad B = \left[\begin{array}{c} B_1 \\ \hline B_2 \end{array}\right],$$

$$F = \left[\begin{array}{c} 0 \\ \hline F_2 \end{array}\right], \quad F_2 = \left[\begin{array}{c} 0 \\ \hline F_0 \end{array}\right]$$

$$C = \left[\begin{array}{c|c} 0 & T \end{array}\right], \quad T^T T = I, \tag{1.4}$$

where $A_{11} \in \mathbb{R}^{(n-p)\times(n-p)}$, $T \in \mathbb{R}^{p\times p}$ is orthogonal, $B_2 \in \mathbb{R}^{p\times m}$, and $F_2 \in \mathbb{R}^{p\times q}$. Matrix $F_0 \in \mathbb{R}^{q\times q}$ is nonsingular.

Now, let us define the τ-delayed variables:

$$\begin{aligned} x_\tau(t) &= x(t-\tau(t)), \;\; f_\tau(t) = f(t-\tau(t)), \\ u_\tau(t) &= u(t-\tau(t)). \end{aligned} \tag{1.5}$$

Then, applying the chain rule for derivatives, system (1.1) can be described using the delayed state variables as follows:

$$\begin{cases} \dot{x}_\tau(t) = (1-\dot{\tau}(t))\left[Ax_\tau(t) + Bu_\tau(t) + Ff_\tau(t)\right] \\ y(t) \;\; = Cx_\tau(t). \end{cases} \tag{1.6}$$

Notice that the time-shifted τ-delayed system has no (explicit) output delay.

The *objective* is to design an observer for the time-shifted system (1.6) that is capable of reproducing the τ-delayed disturbance signal $f_\tau(t)$ from the output signal $y(t)$ and from the input signal $u(t)$. To this end, a generalization of the existing sliding-mode observers for time delays will be proposed.

Remark 1.1 The time-shift approach can be applied to standard nonlinear systems used in the literature of time-delay systems [1, 9], broadening its applicability. As an example, consider the system:

$$\begin{cases} \dot{x}(t) = Ax(t) + \phi(x(t)) + Bu(t) + Ff(t) \\ y(t) = Cx(t-\tau(t)), \end{cases}$$

where the known function $\phi(x(t))$ is Lipschitz in $x(t)$. By applying the method above, it is transformed into

$$\begin{cases} \dot{x}_\tau(t) = (1-\dot{\tau}(t))\left[Ax_\tau(t) + \phi(x_\tau(t)) \right. \\ \qquad\qquad \left. + Bu_\tau(t) + Ff_\tau(t)\right] \\ y(t) \;\; = Cx_\tau(t). \end{cases} \tag{1.7}$$

Notice that it would be very simple to extend all the results presented in this chapter to obtain an appropriate SMO for fault reconstruction in (1.7).

1.1.3 Known Time Delays

In this section, we consider that the measurement delay $\tau(t)$ is known and, based on assumption **(A6)**, that the delayed input $u_\tau(t) = u(t-\tau(t))$ is available.

Consider the following sliding-mode observer:

$$\begin{cases} \dot{\hat{x}}_\tau(t) = (1-\dot{\tau}(t))\left[A\hat{x}_\tau(t) + Bu_\tau(t) - G_l e_y + G_n \nu\right] \\ \hat{y}(t) = C\hat{x}_\tau(t), \end{cases} \tag{1.8}$$

where the output observation error $e_y(t)$ and the state observation error $e(t)$ are defined as

$$e(t) \triangleq \hat{x}_\tau(t) - x_\tau(t) \tag{1.9}$$

$$e_y(t) \triangleq \hat{y}(t) - y(t) = Ce(t). \tag{1.10}$$

From (1.6), (1.8), and (1.9), it can be seen that the observation error $e(t)$ is governed by

$$\dot{e}(t) = (1 - \dot{\tau}(t)) \left[(A - G_l C)e(t) + G_n \nu - F f_\tau(t)\right] \tag{1.11}$$

$$e_y(t) = Ce(t), \tag{1.12}$$

where G_l is chosen such that $A - G_l C$ is Hurwitz.

The discontinuous injection term ν is given by

$$\nu = -\rho \frac{P_0 e_y}{\|P_0 e_y\|}, \tag{1.13}$$

where $P_0 > 0$ is a design matrix. The modulation function or gain ρ is designed so that a sliding mode $e_y(t) \equiv 0$ is reached in finite time.

Remark 1.2 When the measurement delay $\tau(t)$ is constant ($\dot{\tau}(t) = 0$), equations (1.8) and (1.11) for the observer proposed here reduce to those of [12] originally designed for fault reconstruction free of delays, but now estimating the delayed states rather than the current ones.

Matrix G_n is chosen to have the structure:

$$G_n = \begin{bmatrix} -L \\ I_{p \times p} \end{bmatrix} T^T, \tag{1.14}$$

where T is orthogonal as defined in (1.4). Matrix $L = [L_1 \quad 0] \in \mathbb{R}^{(n-p)\times p}$, with $L_1 \in \mathbb{R}^{(n-p)\times(p-q)}$, is chosen such that $LF_2 = 0$ and $A_{11} + LA_{21}$ is Hurwitz, due to assumption **(A4)** (see [2, Lemma 4.3]), for $p < q$.

Remark 1.3 Note that if $p = q$, it will not be possible to determine matrix L. In this case, it is necessary to consider $\det(CF) \neq 0$ in assumption **(A4)**. In this case, the system (A, F, C) is minimum phase and relative degree one, and there exist exactly $(n - p)$ invariant zeros, so that A_{11} is Hurwitz.

By a change of coordinates $e \mapsto \bar{e} = T_L e$, where T_L is a nonsingular linear transformation defined as

$$T_L = \begin{bmatrix} I_{n-p} & L \\ 0 & T \end{bmatrix}, \quad \bar{e} = \begin{bmatrix} \bar{e}_1 \\ \bar{e}_2 \end{bmatrix}, \quad e_y = \bar{e}_2, \tag{1.15}$$

the error dynamics can be described as

$$\begin{cases} \dot{\bar{e}}_1(t) = (1-\dot{\tau}(t))\left(\mathcal{A}_{11}\bar{e}_1(t) + (\mathcal{A}_{12} - \mathcal{G}_{l,1})e_y(t)\right) \\ \dot{e}_y(t) = (1-\dot{\tau}(t))\left(\mathcal{A}_{21}\bar{e}_1(t) + (\mathcal{A}_{22} - \mathcal{G}_{l,2})e_y(t)\right. \\ \qquad\qquad \left. +\nu - \mathcal{F}_2 f_\tau(t)\right) \end{cases}, \tag{1.16}$$

where matrices $(\mathcal{A}, C, \mathcal{F}, \mathcal{G}_l)$ correspond to the transformed matrices (A, C, F, G_l). Matrices $\mathcal{G}_{l,1}$ and $\mathcal{G}_{l,2}$ are partitions of the transformed matrix $\mathcal{G}_l = T_L G_l$ with the appropriate dimensions. The partitions of $\mathcal{A}$ are given by

$$\mathcal{A}_{11} = A_{11} + LA_{21}, \quad \mathcal{A}_{12} = (A_{12} + LA_{22} - A_{11}L - LA_{21})T^T,$$
$$\mathcal{A}_{21} = TA_{21}, \quad \mathcal{A}_{22} = TA_{22}T^T - TA_{21}LT^T.$$

It can be seen that the sliding mode is governed by the eigenvalues of $\mathcal{A}_{11} = A_{11} + LA_{21}$, which is Hurwitz by design. Choosing $\mathcal{G}_{l,1} = \mathcal{A}_{12}$ and $(\mathcal{A}_{22} - \mathcal{G}_{l,2}) = \mathcal{A}_{22}^*$, where $\mathcal{A}_{22}^* \in \mathbb{R}^{p\times p}$ is any stable design matrix, the system (1.16) becomes triangular. Consequently, the eigenvalues of $A - G_l C$ will be those of $\mathcal{A}_{11}$ and $\mathcal{A}_{22}^*$. In the coordinate system of (1.4), G_l is then given by

$$G_l = T_L^{-1}\begin{bmatrix} \mathcal{A}_{12} \\ \mathcal{A}_{22} - \mathcal{A}_{22}^* \end{bmatrix}. \tag{1.17}$$

The symmetric design matrix $P_0 > 0$ of (1.13) must be selected satisfying $(\mathcal{A}_{22}^*)^T P_0 + P_0(\mathcal{A}_{22}^*) < 0$. Also, there exists a matrix $P > 0$, satisfying $(A - G_l C)^T P + P(A - G_l C) < 0$, with the following structure [2]:

$$P = T_L^T \begin{bmatrix} P_1 & 0 \\ 0 & P_0 \end{bmatrix} T_L. \tag{1.18}$$

The convergence of the observer (1.8) to the τ-delayed system (1.6) can be proved in the next theorem.

Theorem 1.1 *Consider the τ-delayed system (1.6) that satisfies* **(A1)** *to* **(A6)** *and its observer (1.8), with (1.13), matrices (1.14) and (1.17), respectively. If the modulation function ρ of the discontinuous term (1.13) satisfies*

$$\rho \geq \alpha\|CF\| + \eta_0, \tag{1.19}$$

where η_0 is any positive scalar, then the state estimation error $e(t)$ converges exponentially to zero. The output error $e_y(t)$ reaches zero in finite time $t = t_s$, and a sliding mode starts on $\mathcal{S} = \{e : e_y = Ce = 0\}$, $\forall t \geq t_s$.

Proof As for Proposition 4.1 in [2], where systems without delay were considered, the proof is based on the Lyapunov candidates $V_1(e) = e^T Pe$, with $e(t)$

from (1.11), for the state stability and convergence, and $V_2(e_y) = e_y^T P_0 e_y$, with $e_y(t)$ from (1.16), for sliding-mode reachability. The only difference is the scalar factor $(1 - \dot{\tau}(t))$, present in their time derivatives. Since such factor is always finite, positive, and uniformly away from zero, due to assumption **(A2)**, the stability and convergence can be concluded from the Lyapunov analysis. □

Once the sliding mode is reached in finite time, then $e_y(t) \equiv \dot{e}_y(t) \equiv 0$, and from (1.16), the error $\bar{e}_1(t)$ will be governed by

$$\begin{cases} \dot{\bar{e}}_1(t) = (1 - \dot{\tau}(t)) \left(\mathcal{A}_{11} \bar{e}_1(t) \right) \\ 0 \quad\;\; = (1 - \dot{\tau}(t)) \left(\mathcal{A}_{21} \bar{e}_1(t) + \nu_{eq} - \mathcal{F}_2 f_\tau(t) \right), \end{cases} \tag{1.20}$$

where ν_{eq} is the equivalent output error injection signal (in the sense of equivalent control [35]), which represents the continuous control signal needed to maintain the sliding mode.

Corollary 1.1 *Under the conditions in Theorem 1.1, the delayed disturbance signal $f_\tau(t)$ in (1.5) can be reconstructed, with arbitrary accuracy, as $\delta \to 0$, by the expression*

$$\hat{f}_\tau(t) = \mathcal{F}_2^\dagger \nu_\delta, \tag{1.21}$$

where ν_δ is the continuous approximation of ν_{eq} defined as

$$\nu_\delta = -\rho \frac{P_0 e_y}{\| P_0 e_y \| + \delta}, \tag{1.22}$$

for a small-scalar parameter $\delta > 0$, and $\mathcal{F}_2^\dagger$ is the left pseudo-inverse of $\mathcal{F}_2 = T_L F_2$, given by

$$\mathcal{F}_2^\dagger \triangleq (\mathcal{F}_2^T \mathcal{F}_2)^{-1} \mathcal{F}_2^T. \tag{1.23}$$

Proof Since, by assumption **(A5)**, the matrix F is full rank, the Moore–Penrose left pseudo-inverse for $\mathcal{F}_2$ exists as defined by (1.23). Therefore, the estimate $\hat{f}_\tau$ of the τ-delayed fault signal $f_\tau(t)$ is given by the expression (1.21). From Theorem 1.1, the sliding mode $e_y \equiv 0$ is reached in finite time and the state observation error is $\bar{e}_1(t) \to 0$ exponentially, and according to (1.20), one has

$$\nu_{eq} \to \mathcal{F}_2 f_\tau(t). \tag{1.24}$$

The equivalent output error injection ν_{eq} can be recovered by using the approximation ν_δ given by expression (1.22), with small enough choice of δ, to achieve any degree of accuracy [4]. □

The main advantage of the above time-shift approach is the possibility of determining the observer gains without appealing to specific methods for time-delay systems, like *Lyapunov–Krasovskii* (LK) or *Lyapunov–Razumikhin* (LR) approaches [14, 16]. From the analytical viewpoint, both LK or LR approaches can lead to conservative results, which may restrict the maximum delay duration to some sufficiently small value. LR is well known to be generally more conservative than LK [15]. On the other hand, to reduce the conservativeness of LK, increased complexity of the proper functionals is required. Obtaining the LMI associated to such functionals involves rather specialized mathematical tools [25, 29]. Moreover, from the numerical viewpoint, the number of the LMI decision variables may greatly increase making the method less attractive [30]. In the present approach, an ideal sliding mode can theoretically be produced for arbitrarily long delays, which in practice also avoids low-frequency chattering problems. The price to be paid is that the fault is identified with some delay. Thus a residual error exists for the estimated and the actual fault signal. It should be stressed that the LR and LK approaches also lead to residual errors and do not identify exactly the fault signal. In [18], the observer is designed with delayed output, according to the original (nontime-shifted) system, and the discontinuous term depends on the delayed error signal. An ideal sliding mode is not achievable, and the reconstruction is subject to chattering effects (see [18, Remark 3]).

The problem of noise was already considered by Edwards et al. [12]. They suggest that the equivalent control, which is used in the fault reconstruction, should be passed by a low-pass filter so that the true fault signal, usually of low frequency, could be extracted or separated from the high-frequency noise signal. We have verified the effectiveness of this measure in our method, provided a time separation is possible, *i.e.*, when the noise does not contain large enough low-frequency components leading to false alarms.

1.1.4 Parametric and Time-Delay Uncertainties

Full knowledge of the plant parameters as well of the time delay $\tau(t)$ is certainly unrealistic. Hence, the accommodation of some uncertainty of the plant characteristics is pertinent. To this end, we show that such uncertainties can be lumped as an additional disturbance, thus reducing the fault reconstruction problem to a well-known formulation [34]. First, let the delay be modeled by

$$\tau(t) = \hat{\tau} + \tau_v(t), \tag{1.25}$$

where $\hat{\tau}$ is some known constant estimate for the output delay and $\tau_v(t)$ is the time-varying delay uncertainty. In addition, let the parametric uncertainties be represented by the bounded lumped signal $D\varsigma(t)$, where $\varsigma \in \mathbb{R}^l$, and $D \in \mathbb{R}^{n\times l}$

is some distribution matrix. Then, system (1.1) becomes

$$\begin{cases} \dot{x}(t) = Ax(t) + Bu(t) + Ff(t) + D\varsigma(t) \\ y(t) = Cx(t - \tau(t)) \end{cases}. \tag{1.26}$$

Applying the time shift presented in Sect. 1.1.2, the delayed system can be described as (see (1.5))

$$\begin{cases} \dot{x}_\tau(t) = (1 - \dot{\tau}(t))\,[Ax_\tau(t) + Bu_\tau(t) \\ \qquad\qquad +Ff_\tau(t) + D\varsigma_\tau(t)] \\ y(t) \;\; = Cx_\tau(t), \end{cases} \tag{1.27}$$

where $\varsigma_\tau(t) = \varsigma(t - \tau(t))$.

Since $\dot{\tau}(t)$ is unknown, the observer considers only some constant (nominal) estimate of the delay $\hat{\tau}$ and the estimate of the delayed input signal $\hat{u}(t) = u(t - \hat{\tau})$:

$$\begin{cases} \dot{\hat{x}}_\tau(t) = A\hat{x}_\tau(t) + B\hat{u}(t) - G_l e_y + G_n \nu \\ \hat{y}(t) \;\; = C\hat{x}_\tau(t), \end{cases} \tag{1.28}$$

where $e_y(t)$ is defined in (1.10) and ν in (1.13). Matrices G_n and G_l are as determined in (1.14) and (1.17), respectively.

The state observation error then presents the following dynamics:

$$\begin{aligned} \dot{e}(t) &= (A - G_l C)e(t) + G_n \nu - Ff_\tau(t) + B\tilde{u}(t) - D\varsigma_\tau(t) \\ &\quad + \dot{\tau}(t)\,[Ax_\tau(t) + Bu_\tau(t) + Ff_\tau(t) + D\varsigma_\tau(t)]\,, \end{aligned} \tag{1.29}$$

where $f_\tau = f(t - \tau(t))$, and $\tilde{u}(t) = \hat{u}(t) - u_\tau(t)$.

It can be seen in (1.29) that delay uncertainty terms in $u(t)$, $u_\tau(t)$, and $x_\tau(t)$ are present in the error dynamics. In order to limit their influence in the design of the modulation function ρ and in the performance of the observer, we assume:

(A7) The system is stable or already stabilized, so that matrix A is Hurwitz.

(A8) The signal $u(t)$ represents an external known (measurable) disturbance. Moreover, it is norm bounded by $\|u(t)\| \leq \beta_u$, for some constant $\beta_u > 0$, $\forall t$.

For the estimation error $\tilde{u}(t)$, from **(A8)**, we can also assume:

(A9) The input error $\tilde{u}(t)$ is zero if the exact delay is known and increases with growing uncertainty $|\tau_v(t)| = |\hat{\tau} - \tau| < \epsilon_v$. Therefore, it can be assumed that a class $\mathcal{K}$ function [23] $\beta_v(\epsilon_v)$ exists such that $\|\tilde{u}(t)\| < \beta_v(\epsilon_v)$.

Remark 1.4 The conditions presented in **(A7)** may arise when the system is stabilized locally but monitored in a remote location subject to delays, where the fault reconstruction is to take place. This is the price of considering uncertain and arbitrarily long delays. As discussed in the time-delays literature, if unstable systems are assumed, the uncertain delays must be sufficiently small so that the delayed term $x_\tau(t)$ can be properly compensated in the control design.

Let us denote $M = \sqrt{\frac{\lambda_{\max}(P_x)}{\lambda_{\min}(P_x)}}$, $\kappa_1 = ||P_x B||$, $\kappa_2 = ||P_x F||$, and $\kappa_3 = ||P_x D||$, where $P_x > 0$ is a symmetric matrix used in the Lyapunov function $V_x(x) = x^T P_x x$. Based on assumption **(A7)**, system (1.26) is BIBS (bounded-input, bounded-state) stable. From assumptions **(A3)**, **(A8)** and reminding that $\|\varsigma(t)\| \leq \beta_\varsigma$, it can be seen that $x(t)$ is ultimately bounded by

$$\|x\| \leq M c_\epsilon, \tag{1.30}$$

where $c_\epsilon = 2(\kappa_1\beta_u + \kappa_2\alpha + \kappa_3\beta_\varsigma) + \epsilon$, for any arbitrarily small $\epsilon > 0$. Since $x(t)$ is ultimately bounded, and considering the bound r_d for $\dot{\tau}(t)$ in **(A2)**, it is possible to aggregate the three last terms of (1.29) as a norm bounded signal $\xi(t)$, $\|\xi(t)\| \leq \bar{\xi}$, such that:

$$\begin{aligned}\xi(t) &\triangleq B\tilde{u}(t) - D\varsigma_\tau(t) + \dot{\tau}(t)[Ax_\tau(t) + Bu_\tau(t) \\ &\quad + Ff_\tau(t) + D\varsigma_\tau(t)],\end{aligned} \tag{1.31}$$

$$\begin{aligned}\bar{\xi} &\triangleq \|B\|\beta_v(\epsilon_v) + \frac{r_d\|B\|\beta_u}{2} + 2r_d\|A\|Mc_\epsilon \\ &\quad + r_d\|F\|\alpha + (1 + r_d)\|D\|\beta_\varsigma.\end{aligned} \tag{1.32}$$

The partition of the uncertainty vector as $\xi = [\xi_1^T \ \ \xi_2^T]$ is made according to the partition of the error $\bar{e} = [\bar{e}_1^T \ \ \bar{e}_2^T]$. Hence, after the change of coordinates (1.15), the equations for the error dynamics become

$$\begin{cases}\dot{\bar{e}}_1(t) &= \mathcal{A}_{11}\bar{e}_1(t) + (\mathcal{A}_{12} - \mathcal{G}_{l,1})\bar{e}_y(t) \\ &\quad +(\xi_1(t) + L\xi_2(t)), \\ \dot{e}_y(t) &= \mathcal{A}_{21}\bar{e}_1(t) + (\mathcal{A}_{22} - \mathcal{G}_{l,2})e_y(t) + \nu \\ &\quad -\mathcal{F}_2 f_\tau(t) + T\xi_2(t).\end{cases} \tag{1.33}$$

Then, the next theorem can be stated:

Theorem 1.2 *Consider the uncertain τ-delayed system (1.27), satisfying assumptions* ***(A1)*** *to* ***(A9)****, and its observer (1.28), where (1.14) and G_l is given by (1.17). If the modulation function ρ of the discontinuous term (1.13) is chosen as*

$$\rho \geq 2\frac{\|TA_{21}\|\mu_0}{\|P\|}\bar{\xi} + \|TF_2\|\alpha + \|TM_2\|\bar{\xi} + \eta_0, \tag{1.34}$$

where $M_2 = [0^{p\times(n-p)} \ \ I^{p\times p}]$, $\mu_0 = -\lambda_{\max}((A - G_lC)^T P + P(A - G_lC))$, with $P > 0$ as in (1.18), $\bar{\xi}$ is given in (1.32), and η_0 is any positive scalar, then the state estimation error $e(t)$ is ultimately bounded, and the output error $e_y(t)$ reaches *zero in finite time $t = t_s$ so that a sliding mode starts on $\mathscr{S} = \{e : e_y = Ce = 0\}$, $\forall t \geq t_s$.*

Proof Substituting (1.31) in (1.29), one has

$$\dot{e}(t) = (A - G_l C)e(t) + G_n \nu - F f_\tau(t) + \xi(t). \tag{1.35}$$

Since the unknown terms in (1.35) were grouped into a bounded signal $\xi(t)$, it is only needed to follow the steps of [34, Lemma 1 and Proposition 2], using Lyapunov candidates $V_3(e) = e^T P e$, with $e(t)$ from (1.35), for error boundedness, and $V_4(e_y) = e_y^T P_0 e_y$, with $e_y(t)$ from (1.33), for sliding-mode reachability to conclude this proof. □

Observe that Theorem 1.2 only requires the norm bound $\bar{\xi}$ from (1.32) to determine the gain from the discontinuous term. Hence, this observer design does not require the complete knowledge of the delay, only its duration estimate $\hat{\tau}$ from (1.25) and the norm bound for the delay rate r_d from assumption **(A2)**.

On the sliding mode $e_y(t) \equiv 0$, error equation (1.33) becomes

$$\begin{cases} \dot{\bar{e}}_1(t) &= \mathcal{A}_{11}\bar{e}_1(t) + (\xi_1(t) + L\xi_2(t)), \\ 0 &= \mathcal{A}_{21}\bar{e}_1(t) + \nu_{eq} - \mathcal{F}_2 f_\tau(t) + T\xi_2(t). \end{cases} \tag{1.36}$$

In contrast to the previous case, where $\xi(t) = 0$, here, the error $\bar{e}_1$ does not converge to zero but depends on the uncertainty term $\xi(t)$. However, its effect can be minimized according to the following procedure.

Consider a scaling matrix $W \triangleq \begin{bmatrix} W_1 & F_0^{-1} \end{bmatrix}$, where $W_1 \in \mathbb{R}^{q\times(p-q)}$ is a design matrix, and F_0 is the lower block of the matrix F as in (1.4). The matrix W will be used to reconstruct the fault signal using the expression:

$$\hat{f}_\tau(t) = WT^T \nu_{eq}. \tag{1.37}$$

Multiplying the second equation of (1.36) by W and rearranging its terms, we obtain

$$\begin{cases} \dot{\bar{e}}_1(t) &= \mathcal{A}_{11}\bar{e}_1(t) + (\xi_1(t) + L\xi_2(t)) \\ \hat{f}_\tau(t) &= -WT^T\mathcal{A}_{21}\bar{e}_1(t) - W\xi_2(t) + f_\tau(t) \end{cases}. \tag{1.38}$$

Therefore, the estimated value for the fault signal is given by

$$\hat{f}_\tau(t) = f_\tau(t) + G(s)\xi(t), \tag{1.39}$$

where $G(s)$ is a transfer function between the signal $\xi(t)$ and the estimated fault signal, and it is defined as

$$G(s) = -\, WT^T\mathcal{A}_{21}\left[(sI - \mathcal{A}_{11})^{-1}(M_1 + LM_2)\right] - WQ_2, \tag{1.40}$$

where $M_1 = [I^{(n-p)\times(n-p)} \quad 0^{(n-p)\times p}]$ and $M_2 = [0^{p\times(n-p)} \quad I^{p\times p}]$. The gain of the transfer function $G(s)$ can then be minimized by using the *bounded real lemma* [7] to find the appropriate scale matrix W.

1.1.5 Simulation Example

Consider the example proposed in [12], an inverted pendulum with a cart, linearized around the equilibrium point. The system matrices for the state-space representation already transformed into the canonical form (1.4) are

$$A = \begin{bmatrix} -0.1738 & 0 & 36.9771 & -6.2589 \\ 0 & 0 & 0 & 1 \\ 1 & 0 & 0 & 0 \\ -0.0091 & 0 & 1.9333 & -1.9872 \end{bmatrix},$$

$$B = \begin{bmatrix} -1.0095 \\ 0 \\ 0 \\ -0.3205 \end{bmatrix}, \; F = \begin{bmatrix} 0 \\ 0 \\ 0 \\ -0.3205 \end{bmatrix}, \; C = \begin{bmatrix} 0 & -1 & 0 & 0 \\ 0 & 0 & 1 & 0 \\ 0 & 0 & 0 & -1 \end{bmatrix}. \tag{1.41}$$

Assume that an input disturbance is given by u(t)= 5 sin(0.5t), which is assumed known but, due to the delay uncertainty, generates the input uncertainty term $\tilde{u}(t)$ defined by (26). Then, the fault reconstruction observer is designed as follows:

1. Compute matrix $L = [L_1 \quad 0] \in \mathbb{R}^{(n-p)\times p}$, with $L_1 \in \mathbb{R}^{(n-p)\times(p-q)}$, such that $LF_2 = 0$ and $A_{11} + LA_{21}$ is Hurwitz. The solution is nonunique. We have set $L = [0 \; -9.83 \; 0]$, which trivially makes $A_{11} + LA_{21} = [-10]$.
2. Using L and T (see (1.4)), compute T_L in (1.15) and obtain the transformed matrix $\mathcal{A} = T_L A T_L^{-1}$ and its partitions.
3. Choose the stable design matrix $\mathcal{A}_{22}^* \in \mathbb{R}^{(p\times p)}$ and compute matrix G_l in (1.17) using T_L the partitions of $\mathcal{A}$. This solution is also nonunique. We have set $\mathcal{A}_{22}^* = \text{diag}(-15, -16, -17)$ to allocate the poles of $A - G_l C$ in $\{-10, -15, -16, -17\}$.
4. Then, from (1.14) and (1.17), we obtain

$$G_n = \begin{bmatrix} 0 & 9.83 & 0 \\ -1 & 0 & 0 \\ 0 & 1 & 0 \\ 0 & 0 & -1 \end{bmatrix}, \; G_l = \begin{bmatrix} 0 & 192.49 & 6.29 \\ -15 & 0 & -1 \\ 0 & 25.82 & 0 \\ 0 & 1.84 & 1-5.01 \end{bmatrix}. \tag{1.42}$$

5. Choose P_0 satisfying $(\mathcal{A}_{22}^*)^T P_0 + P_0(\mathcal{A}_{22}^*) < 0$. Trivially, $P_0 = I$.
6. Determine ρ satisfying (1.34). In our case, a constant $\rho = -3$ was used.

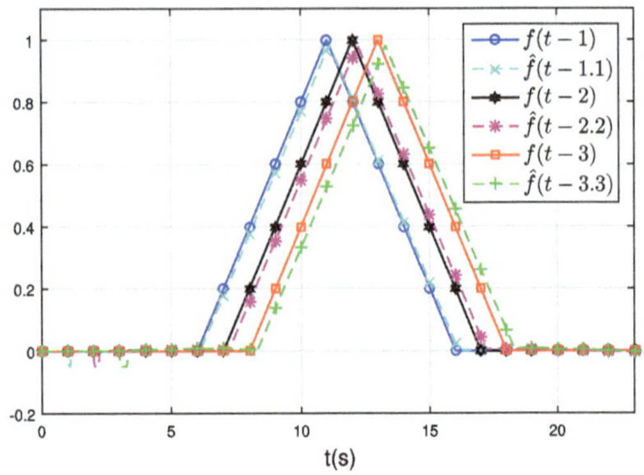

Fig. 1.1 Fault signal reconstruction for several measurement delay lengths, with 10% uncertainty on the duration

7. Determine the scaling matrix $W = [W_{1,1} \quad W_{1,2} \quad F_0^{-1}]$ to be used in the reconstruction equation (1.37). Since $\xi(t) = B\tilde{u}(t)$, Eq. (1.39) can be written as $\hat{f}_\tau(t) = f_\tau(t) + G_u(s)\tilde{u}(t)$, where $G_u(s) = G(s)B$. Substituting values, we obtain

$$G_u(s) = -\frac{s - 1.01W_{1,2} + 9.97}{s + 10}. \tag{1.43}$$

From this transfer function, it can be seen that $W_{1,1}$ does not influence $G_u(s)$; hence, it was set to zero. The value of $W_{1,2}$ was manually adjusted to $W_{1,2} = 7.5$ using the bode plot of $G_u(s)$ so that the matrix $W = \begin{bmatrix} 0 \; 7.5 \; -3.12 \end{bmatrix}$ minimizes the reconstruction error of the fault signal to less than 1%.

In order to simulate the delay uncertainty, a 10% difference between the true and the nominal delay was introduced, i.e., $\tau = 1.1\hat{\tau}$. Figure 1.1 presents the simulation results for reconstructing a triangular signal representing an actuator fault. The current delays were 1.1s, 2.2s, and 3.3s, while the nominal delays to design the observer were 1s, 2s, and 3s, respectively.

1.2 Sampled-Output Systems

This section presents the application of the proposed approach to develop a fault reconstruction method for the specific case of systems with sampled output. We consider sampled-output systems with constant sampling period h that occur in practice when the output signal measurement is not continuously available.

In [17], the sampled output is reconstructed by a zero-order hold (ZOH) that is modeled as a time-varying output delay. In contrast, here the idea is to reconstruct the continuous-time signal that generates the sampled output, departing from the latter. Then, the original sampled-output system can be viewed as an unsampled continuous-time system with *constant* delay of one sampling time. To improve the reconstruction quality, one can wait for more samples and then use more effective interpolation methods. This is equivalent to *deliberately* reducing the problem to a system with constant output delays of $n_s \in \mathbb{N}$ sampling periods in the output signal, where n_s is the number of samples. Then, the time-shift method for observer design of Sect. 2.3 is applied to the reconstructed system.

1.2.1 Problem Statement

Consider a linear time-invariant system with sampled output:

$$\begin{cases} \dot{x}(t) = Ax(t) + Bu(t) + Ff(t) \\ y(t) = Cx(t) \\ y_k = y(kh) \end{cases}, \tag{1.44}$$

where $k \in \mathbb{N}$ is the sample index, and h is the sampling period. The output signal $y(t) \in \mathbb{R}^p$ is not directly available, only its sampled version y_k, which is only valid at instants $t = kh$. The input signal $u(t) \in \mathbb{R}^m$ is measurable, and the system states $x(t) \in \mathbb{R}^n$ are not measurable. The unknown disturbance signal $f(t) \in \mathbb{R}^p$, $q \leq p < n$ represents the effect of an actuator fault. Consider that assumptions **(A1)**–**(A6)** are still valid.

Then, the objective is to estimate the continuous-time fault signal $f(t)$ from the measurable signals and the plant model.

1.2.2 Recovering the Continuous-Time Signal from Its Sampled Values

A well-known reconstruction of the continuous-time output $y(t)$ is the ZOH (zero-order hold) filter. As a rule, the reconstructed output will present an approximation error. For the observers, this would appear as an output measurement disturbance. Even though SMO have some tolerance to disturbances, output disturbance can strongly deteriorate its performance, and causal filters like ZOH may lead to unacceptable reconstruction error for SMO. Therefore, a suitable reconstruction filter must be chosen so that the approximation error is small enough. Interpolation reconstruction filters yield smaller reconstruction errors [8], but have the disadvantage of being noncausal.

Our strategy, as outlined above, is to intentionally allowing, say, $n_s \in \mathbb{N}$ sampling periods of delay in the reconstructed output signal y_R given by

$$y_R(t) = Cx(t - n_s h) + e_r(t), \tag{1.45}$$

where e_r is the reconstruction error.

From (1.44) and (1.45), the resulting system with continuous-time reconstructed output is given by

$$\begin{cases} \dot{x}(t) = Ax(t) + Bu(t) + Ff(t) \\ y_R(t) = Cx(t - n_s h) + e_r(t). \end{cases} \tag{1.46}$$

Since the output delay now is constant and known, assumption **(A2)** is also valid with $\tau = n_s h$ and $\dot{\tau}(t) = 0$, and the time-shift method presented in Sect. 1.1.2 can be directly applied to reconstruct the delayed disturbance signal $f_\tau \triangleq f(t - n_s h)$. To this end, the SMO is given by

$$\begin{cases} \dot{\hat{x}}_\tau(t) &= A\hat{x}_\tau(t) + Bu_\tau(t) - G_l e_y + G_n \nu \\ e_y(t) &= \hat{y}_R(t) - y_R(t) \\ \hat{y}_R(t) &= C\hat{x}_\tau(t) \end{cases} , \tag{1.47}$$

where $\hat{x}_\tau(t) = \hat{x}(t - n_s h)$ is the estimate of the delayed state and $u_\tau(t) = u(t - n_s h)$.

The discontinuous injection term ν is given by

$$\nu = -\rho \frac{P_0 e_y}{\|P_0 e_y\|}. \tag{1.48}$$

The gain ρ is designed so that a sliding mode $e_y(t) \equiv 0$ is reached in finite time. Matrices $P_0 > 0$, G_l, and G_n are designed as in Sect. 1.1.

1.2.3 Sampled-Data Reconstruction Methods

The simplest reconstruction method is the zero-order hold (ZOH), in which the value presented by the system is held constant until a new value is available:

$$y_{ZOH}(t) = y(kh), \quad \forall t \in [kh, (k+1)h). \tag{1.49}$$

Since this filter is causal and of simple implementation, it is commonly used in most applications.

Another causal method is the first-order hold (FOH):

$$y_{FOH}(t) = y(kh) + [y(kh) - y((k-1)h)] \frac{t \quad kh}{h}, \quad \forall t \in [kh, (k+1)h). \tag{1.50}$$

The interpolation first-order hold (IFOH) method can also be used to reconstruct sampled data, by means of the linear interpolation between the current sample and the next one [8]. Since both samples are part of the interpolation, the error is zero at every sample, and there are no discontinuous jumps. On the other hand, because of its dependency from the next sample, the IFOH is a noncausal process. However, if a time delay of one sample is allowed, the IFOH can be implemented as a causal algorithm:

$$y_{IFOH}(t) = y\left((k-1)\,h\right) + \left[y\left(kh\right) - y\left((k-1)\right)h\right]\frac{t-kh}{h},$$
$$\forall t \in \left[kh, (k+1)\,h\right). \quad (1.51)$$

In [37], a second-order hold (SOH) based on Taylor series expansion to increase reconstruction accuracy is presented. This SOH is also extrapolative and presents discontinuous jumps at every new sample.

Here, we propose the use of a interpolation second-order hold (ISOH), by adding a two-sample time delay and using Lagrange's polynomial interpolation [32] to reconstruct the signal. Since all sample points in the interval are used for interpolation, there will be no discontinuous jumps. The expression for ISOH can be written in a matrix form as

$$y_{ISOH}(t) = \begin{bmatrix} 1 & 0 & 0 \\ 1 & h & h^2 \\ 1 & 2h & (2h)^2 \end{bmatrix}^{-1} \cdot \begin{bmatrix} y\left((k-2)\,h\right) \\ y\left((k-1)\,h\right) \\ y\left(kh\right) \end{bmatrix} \cdot \begin{bmatrix} 1 \; t \; t^2 \end{bmatrix},$$
$$\forall t \in \left[kh, (k+2)\,h\right). \quad (1.52)$$

Figure 1.2 presents a sampled sinusoidal signal reconstructed using the interpolation methods compared to the original signal and to the ZOH.

1.2.4 Influence of the Reconstruction Error

From (1.46) and (1.47), the state observation error $e(t) = \hat{x}_\tau(t) - x_\tau(t)$ is governed by

$$\dot{e}(t) = Ae(t) - G_l e_y(t) + G_n v - F f_\tau(t), \quad (1.53)$$

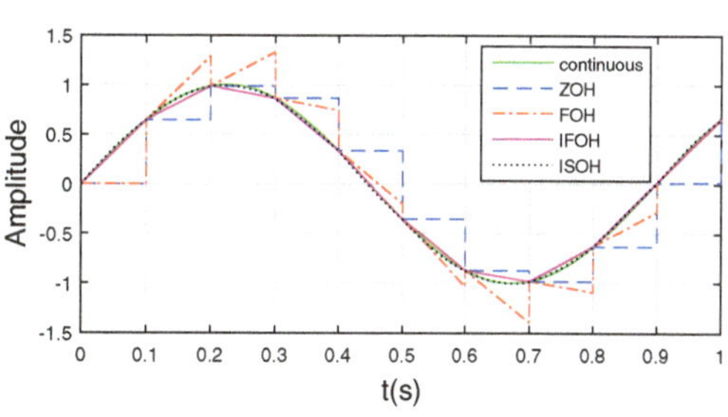

Fig. 1.2 Comparison between ZOH, FOH, IFOH, and ISOH reconstruction methods for the signal $x = \sin(5t)$ sampled at a 0.1 sec period

where $x_\tau(t) = x(t - n_s h)$ and $f_\tau = f(t - n_s h)$ are the delayed state and the delayed fault signal, respectively.

Applying the change of coordinates $\bar{e} = T_L e(t)$ as in (1.15), the error dynamics can be written as

$$\begin{cases} \dot{\bar{e}}_1(t) = \mathcal{A}_{11}\bar{e}_1(t) + \mathcal{A}_{12}\bar{e}_2(t) - \mathcal{G}_{l,1}e_y(t) \\ \dot{\bar{e}}_2(t) = \mathcal{A}_{21}\bar{e}_1(t) + \mathcal{A}_{22}\bar{e}_2(t) - \mathcal{G}_{l,2}e_y(t) \\ \qquad\quad +\nu - \mathcal{F}_2 f_\tau(t). \end{cases} \tag{1.54}$$

From the structures of C in (1.4) and T_L in (1.15), it can be seen that $\bar{e}_2(t) = Ce(t)$. Then, from Eq. (1.47), it can be seen that

$$e_y(t) = Ce(t) - e_r(t) = \bar{e}_2(t) - e_r(t) \tag{1.55}$$

$$\dot{e}_y(t) = \dot{\bar{e}}_2(t) - \dot{e}_r(t). \tag{1.56}$$

Substituting (1.55) and (1.56) into (1.54), one obtains

$$\begin{cases} \dot{\bar{e}}_1(t) \qquad\qquad = \mathcal{A}_{11}\bar{e}_1(t) + \mathcal{A}_{12}(e_y(t) + e_r(t)) \\ \qquad\qquad\qquad\quad -\mathcal{G}_{l,1}e_y(t) \\ \dot{e}_y(t) + \dot{e}_r(t) = \mathcal{A}_{21}\bar{e}_1(t) + \mathcal{A}_{22}(e_y(t) + e_r(t)) \\ \qquad\qquad\qquad\quad -\mathcal{G}_{l,2}e_y(t) + \nu - \mathcal{F}_2 f(t). \end{cases} \tag{1.57}$$

Rearranging terms, the error dynamics are described as

$$\begin{cases} \dot{\bar{e}}_1(t) = \mathcal{A}_{11}\bar{e}_1(t) + (\mathcal{A}_{12} - \mathcal{G}_{l,1})e_y(t) + \mathcal{A}_{12}e_r \\ \dot{e}_y(t) = \mathcal{A}_{21}\bar{e}_1(t) + (\mathcal{A}_{22} - \mathcal{G}_{l,2})e_y(t) + \nu \\ \qquad\quad -\mathcal{F}_2 f(t) + \mathcal{A}_{22}e_r(t) - \dot{e}_r(t). \end{cases} \tag{1.58}$$

To analyze the influence of the reconstruction error, the following assumption will be adopted:

(A10) For the n-order interpolation reconstruction filter, the output $y(t)$ and its $(n+1)$ derivative, $y^{(n+1)}(t)$, are continuous in the interval $t \in [(k-n)h,\ kh]$, and $y^{(n+1)}(t)$ is norm bounded by a known constant $\bar{y}^{(n+1)}$, *i.e.*, $\left\| \frac{d^{(n+1)}y(t)}{dt^{(n+1)}} \right\| \leq \bar{y}^{(n+1)}$.

Then, a bound for the reconstruction error e_r (1.45) for a polynomial interpolation of order n can be determined by Süli and Mayers [32, Theorem 6.2]:

$$\|e_r(t)\| \leq \frac{M_{n+1}}{(n+1)!}\pi_{n+1}(t), \quad \forall t \in [(k-n)h, kh], \tag{1.59}$$

where

$$\pi_{n+1}(t) = (t-kh)(t-(k-1)h)\dots(t-(k-n)h),$$

$$M_{n+1} = \max_{\zeta\in((k-n)h,kh)} \|y^{(n+1)}(\zeta)\|.$$

Also, for $n \geq 1$, a bound for $\dot{e}_r(t)$ can be written as [32, Corollary 6.1]

$$\|\dot{e}_r(t)\| \leq \frac{(nh)^n M_{n+1}}{n!}, \quad \forall t \in [(k-n)h, kh]. \tag{1.60}$$

For the zero-order hold (ZOH), applying $n = 0$, it can be seen that

$$\|e_r(t)\| \leq h \max_{\zeta\in(kh,(k+1)h)} \|\dot{y}(\zeta)\| \leq h\bar{y}^{(1)}. \tag{1.61}$$

The derivative of the reconstruction error $\dot{e}_r$ is not norm bounded in the ZOH because the error shows a discontinuous jump at every new sample.

In the case of the interpolation first-order hold (IFOH), the reconstruction error is given by

$$\|e_r(t)\| \leq \frac{h^2}{8} \max_{\zeta\in((k-1)kh,kh)} \|\ddot{y}(\zeta)\| \leq \frac{h^2}{8}\bar{y}^{(2)}, \tag{1.62}$$

$$\|\dot{e}_r(t)\| \leq h\bar{y}^{(2)}. \tag{1.63}$$

For the interpolation second-order hold (ISOH), we have

$$\|e_r(t)\| \leq \frac{h^3}{9\sqrt{3}} \max_{\zeta\in((k-2)kh,kh)} \|y^{(3)}(\zeta)\| \leq \frac{h^3}{9\sqrt{3}}\bar{y}^{(3)}, \tag{1.64}$$

$$\|\dot{e}_r(t)\| \leq 2h^2\bar{y}^{(3)}. \tag{1.65}$$

Since the relative degree from $f(t)$ to $y(t)$ is one, differentiating the output $y(t)$, an upper bound can be obtained from the norm bound of $f(t)$ in **(A3)**. Consequently, the n-th derivatives of $y(t)$ can be norm bounded by the norm bounds for the $(n-1)$ derivatives of $f(t)$, which exist for sufficiently smooth faults. Considering that the relative degree from $u(t)$ to $y(t)$ is n^*, an upper bound can be obtained from the norm bound for the $\max(0, n - n^*)$ derivatives of $u(t)$, where order 0 means just $u(t)$ bounded.

To determine a modulation function ρ that allows a sliding mode to be reached on the set $\mathcal{S} = \{e_y : e_y = Ce = 0\}$, a first-order approximation filter (FOAF) can be used to estimate an upper bound for the state $\bar{e}_1$ of the error system (1.58).

Under assumption **(A10)**, an instantaneous upper bound for the reconstruction error $\|e_r(t)\| \leq \bar{e}_r$ is known, and from (1.59), its derivative is also bounded, $\|\dot{e}_r(t)\| \leq \bar{\dot{e}}_r$. Let γ_0 be the stability margin of $G_y(s) \triangleq (sI - \mathcal{A}_{11})^{-1}\left(\mathcal{A}_{12} - \mathcal{G}_{l,1}\right)$

and γ_R be the stability margin of $G_R(s) \triangleq (sI - \mathcal{A}_{11})^{-1}\mathcal{A}_{12}$. From **(A8)**, it can be assumed that $\gamma_0 \leq \gamma_R$. Let $\gamma \triangleq \gamma_0 - \delta_0$, with $\delta_0 > 0$ being an arbitrary constant. Applying [20, Lemma 2] to the equation of $\bar{e}_1(t)$ in (1.58), it can be shown that

$$\|\bar{e}_1\| \leq \beta_{e_1} \triangleq e^{-\gamma t} * \left(c_1\|e_y(t)\| + c_2\bar{e}_r(t)\right) \\ + c_3 e^{-(\lambda_0-\delta_0)}\|e_1(0)\|, \tag{1.66}$$

where $\lambda_0 := \min_j\{-Re(\lambda_j)\}$ is the stability margin of $\mathcal{A}_{11}$, and

$$c_1 = \frac{\|P_e(\mathcal{A}_{12} - \mathcal{G}_{l,1})\|}{\lambda_{\max}(P_e)}, \quad c_2 = \frac{\|P_e\mathcal{A}_{12}\|}{\lambda_{\max}(P_e)}, \quad c_3 = \sqrt{\frac{\lambda_{\max}(P_e)}{\lambda_{\min}(P_e)}}, \tag{1.67}$$

where $P_e > 0$ satisfies the Lyapunov equation $(\mathcal{A}_{11}^T P_e + P_e\mathcal{A}_{11}) = -2I$.

Since $\bar{e}_1$ is bounded, a sliding mode $e_y \equiv 0$ can be achieved in finite time. Then a reconstruction of the actuator fault can be obtained, added by a filtered signal from the reconstruction error. This highlights the importance of choosing an appropriate order for the interpolation filter to minimize reconstruction error.

Now, for a sampled-output system (1.44) with sampling period h and the number of samples n_s corresponding to the total time shift $\tau = n_s h$ required by the continuous-time output reconstruction algorithm (see above (1.45)), the following theorem can be stated.

Theorem 1.3 *Consider the observer (1.47) designed to estimate the states of the plant (1.46), satisfying assumptions* **(A1)** *to* **(A6)***, and the observer error dynamics in (1.58). Under assumption* **(A10)***, the reconstruction error is bounded by $\|e_r(t)\| \leq \bar{e}_r$ from (1.59), and its derivative is also bounded by $\|\dot{e}_r(t)\| \leq \bar{\dot{e}}_r$ according to (1.60). Choosing the modulation function of the discontinuous term ν (1.48) of the observer to satisfy*

$$\rho \geq \|\mathcal{A}_{21}\|\beta_{e_1} + \|\mathcal{A}_{22}\|\bar{e}_r + \|\mathcal{F}_2\|\alpha + \bar{\dot{e}}_r + \eta, \tag{1.68}$$

modulo exponentially decaying terms due to the transient terms of the FOAFs in (67), and $\eta > 0$ being arbitrary constant, then an ideal sliding mode will be reached on the set $\mathcal{S} = \left\{e_y \in \mathbb{R}^p : e_y = 0\right\}$ in finite time, where $e_y(t)$ is given by (1.47). Moreover, on the sliding mode, the delayed fault $f_\tau(t) = f(t - n_s h)$ can be estimated by

$$\hat{f}_\tau(t) = \mathcal{F}_2^\dagger \nu_\delta \tag{1.69}$$

as in Corollary 1.1. Such estimate obeys, with arbitrary accuracy as $\delta \to 0$, the expression:

$$\hat{f}_\tau(t) = f_\tau(t) - \mathcal{F}_2^\dagger\left(\mathcal{A}_{21}\bar{e}_1(t) + \mathcal{A}_{22}e_r - \dot{e}_r\right). \tag{1.70}$$

Proof Consider the Lyapunov candidate function for the output observation error $V_y = e_y^T P_0 e_y$, where $P_0 > 0$ is defined in (1.18). From assumption **(A8)** and considering the structure of P in (1.18), P_0 is a solution for the Lyapunov equation $P_0\left(\mathcal{A}_{22} - \mathcal{G}_{l,2}\right) + \left(\mathcal{A}_{22} - \mathcal{G}_{l,2}\right)^T P_0 = -I$. The derivative of V_y along the trajectories of (1.58) is

$$\dot{V}_y = -e_y^T e_y + 2e_y^T P_0 \left(\mathcal{A}_{21}\bar{e}_1 + \mathcal{A}_{22}e_r + \mathcal{F}_2 f_\tau - \dot{e}_r\right) - 2\rho\|e_y^T P_0\|. \tag{1.71}$$

Applying the bounds β_{e_1} in (1.66), $\bar{e}_r$ in (1.59), and $\dot{\bar{e}}_r$ in (1.60), and α from assumption **(A3)**, the following inequality can be written as

$$\begin{aligned}\dot{V}_y \leq &- \|e_y\|^2 + 2\|P_0 e_y\|\left(\|\mathcal{A}_{21}\|\beta_{e_1} + \|\mathcal{A}_{22}\|\bar{e}_r\right.\\ &\left.+\|\mathcal{F}_2\|\alpha + \dot{\bar{e}}_r - \rho\right).\end{aligned} \tag{1.72}$$

It can be seen that, if ρ is given by (1.68), (1.72) becomes

$$\dot{V}_y \leq -\|e_y\|^2 - 2\eta\|P_0 e_y\|, \tag{1.73}$$

showing that the set $\mathcal{S}$ is reached in finite time.

When the sliding mode $e_y \equiv \dot{e}_y \equiv 0$ is initiated, from (1.58), one has

$$\nu_{eq} = -\mathcal{A}_{21}\bar{e}_1(t) + \mathcal{F}_2 f_\tau(t) - \mathcal{A}_{22}e_r + \dot{e}_r. \tag{1.74}$$

Then according to Corollary 1.1, Sect. 1.1.3, the fault can be estimated from the equivalent control using

$$\hat{f}_\tau(t) = \mathcal{F}_2^\dagger \nu_\delta. \tag{1.75}$$

Rearranging the terms of (1.74), one obtains the expression for the reconstructed fault (1.70), as $\delta \to 0$. □

From this result, it can be seen that a filtered signal from the reconstruction error will additively influence the fault estimate signal. Note that in the sliding mode $e_y \equiv 0$, from (1.58) $\dot{\bar{e}}_1(t) = \mathcal{A}_{11}\bar{e}_1 + \mathcal{A}_{12}e_r(t)$, so $\bar{e}_1$ depends only on the reconstruction error *modulo* an exponentially decaying term.

The reconstruction error e_r and its derivative $\dot{e}_r$ can be made arbitrarily small by choosing a small sampling period h in (1.59) and (1.60). Therefore, it can be seen from (1.70) that $\hat{f}_\tau(t)$ approximates $f_\tau(t)$ as $t \to +\infty$.

Using n_s samples, it is possible to use a polynomial of degree n_s by the same method. In fact, it can also be used with splines or other interpolation methods considering the effect of the reconstruction error in Theorem 1.3.

1.2.5 Numerical Example

Case 1 Consider a system with the same dynamics of the example of Sect. 1.1.5, but now with sampled output as in (1.44). Let the system be disturbed with a sinusoidal fault signal. An observer (1.47) is designed with the same G_l and G_n as in Sect. 1.1.5 since these gains do not depend on the delay duration. A constant ρ is selected so that the inequality (1.68) is satisfied. Numerical simulations were performed for sampling periods of 20 ms, 60 ms, and 100 ms, comparing the performance of the observer for the three reconstruction filters proposed [zero-order hold (ZOH), interpolation first-order hold (IFOH), and interpolation second-order hold (ISOH)]. In order to obtain a quantitative comparison, the root mean-squared error percentage (RMSE%) and the mean absolute error percentage (MAE%) are defined as

$$\text{RMSE\%}_{f_\tau(t)} = \frac{\sqrt{\mathbb{E}\left[(f_\tau(t) - \hat{f}_\tau(t))^2\right]}}{\sqrt{\mathbb{E}\left[f_\tau^2(t)\right]}} \times 100\% \tag{1.76}$$

$$\text{MAE\%}_{f_\tau(t)} = \frac{\mathbb{E}\left[\left|f_\tau(t) - \hat{f}_\tau(t)\right|\right]}{\mathbb{E}\left[|f_\tau(t)|\right]} \times 100\%. \tag{1.77}$$

Table 1.1 shows that the ISOH (second-order filter) presents better performance in reconstructing the fault. Figures 1.3, 1.4, and 1.5 present the fault reconstruction for all developed strategies, assuming 60 ms sampling period.

Case 2 In order to compare the reconstruction method presented in this chapter with the existing literature, a simple system was used. Consider a first-order system:

$$\dot{x}(t) = -5x(t) + u(t) + f(t) \tag{1.78}$$

$$y_k = x(kh) \tag{1.79}$$

with a sampling time $h = 50$ ms, and a fault signal $f(t) = 0.6\sin(2t)$.

Table 1.1 Error metrics for the reconstructed delayed fault in Case 1

Hold type	h(ms)	RMSE%	MAE%
ZOH	20	37.02	30.72
	60	57.03	52.26
	100	71.16	61.60
IFOH	20	2.00	1.78
	60	6.52	6.07
	100	12.51	12.00
ISOH	20	0.21	0.18
	60	1.07	0.94
	100	3.32	3.15

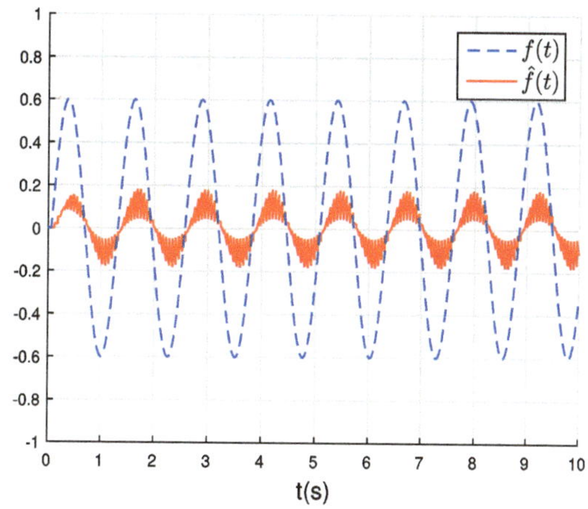

Fig. 1.3 Fault reconstruction of a sampled data system with 60 ms sampling time using a zero-order hold

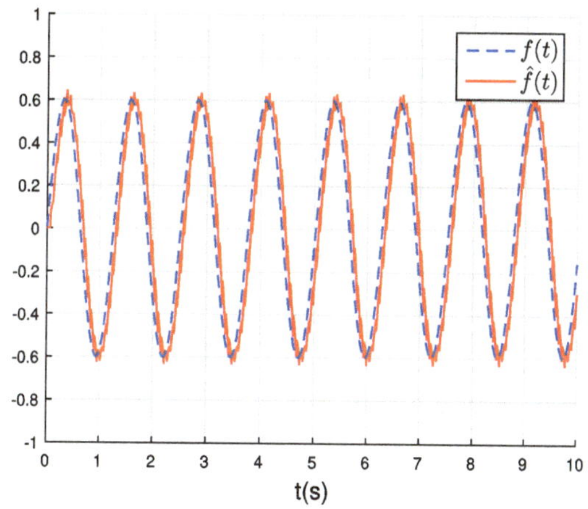

Fig. 1.4 Fault reconstruction of a sampled data system with 60 ms sampling time using an interpolation first-order hold

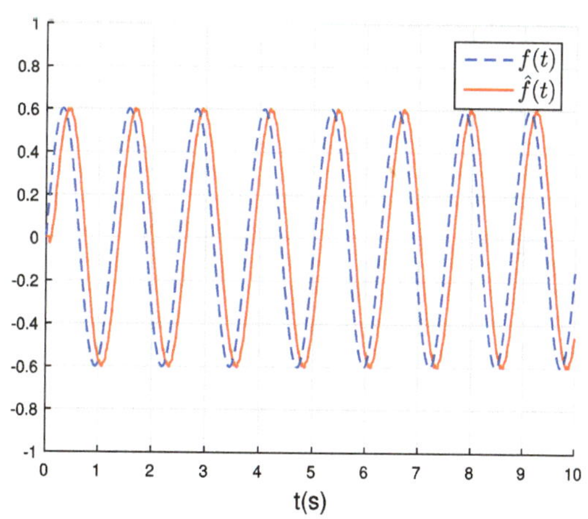

Fig. 1.5 Fault reconstruction of a sampled data system with 60 ms sampling time using an interpolation second-order hold

The fault reconstruction is performed using the method presented in Sect. 1.1.3, with ISOH (1.52) as the output reconstruction filter:

$$\dot{x}(t) = -5x(t) + u(t) + f(t) \tag{1.80}$$

$$y_{ISOH}(t) = x(t-0.1) + e_r(t), \tag{1.81}$$

with $n_s = 2$ and $\tau = 0.1$. The observer is given by

$$\begin{cases} \dot{\hat{x}}_\tau(t) &= -5\hat{x}_\tau(t) + u_\tau(t) - 2e_y + \nu \\ e_y(t) &= \hat{y}_{ISOH}(t) - y_{ISOH}(t) \\ \hat{y}_{ISOH}(t) &= \hat{x}_\tau(t), \end{cases} \tag{1.82}$$

where the discontinuous injection term $\nu = -2e_y/\|e_y\|$. The reconstruction is performed using (1.21), which in this example is given by $\hat{f}_\tau(t) = \nu_\delta$ where $\nu_\delta = -2e_y/\left(\|e_y\| + \delta\right)$, with $\delta = 10^{-4}$. Figure 1.6a presents that the reconstructed signal is continuous; it reproduces closely the fault signal delayed by $\tau = n_s h = 0.1$ after some initial transient.

As a comparison result, an observer in the form proposed in [17] was also tested:

$$\dot{\hat{x}}(t) = -5\hat{x} + \frac{1}{\mu}\left(x(kh) - \hat{x}(kh)\right) - \nu(kh) \tag{1.83}$$

$$\hat{y}_k = x(kh) \tag{1.84}$$

$$\nu(kh) = -M_\beta \text{sign}\left(x(kh) - \hat{x}(kh)\right) \tag{1.85}$$

$$M_\beta = \alpha + \delta_1\alpha \tag{1.86}$$

$$\|f(t)\| \le \alpha, \quad \alpha > 0. \tag{1.87}$$

The sampling kh is implemented as a time-varying delay, where $t - \tau(t) = kh,\ \forall t \in [kh, (k+1)h)$. Using the method in [17], the observer is feasible with $\mu = 0.05$ and $\delta_1 = 0.1$. The fault reconstruction shown in Fig. 1.6b is performed by [17, eq. (31)]:

$$\hat{f}_\tau(t) = \nu_\delta(kh) - \frac{1}{\mu}\left(x(kh) - \hat{x}(kh)\right), \tag{1.88}$$

where $\nu_\delta = -2e_y/\left(\|e_y\| + \delta\right)$, with $\delta = 0.01$. The result presents significant chattering that could only be reduced by removing the discontinuous term $\nu(kh)$, in which case the observer becomes purely linear. The method presented in this chapter manipulates the system so that there is no delay in the observer feedback loop, allowing it to achieve an ideal sliding mode, avoiding chattering.

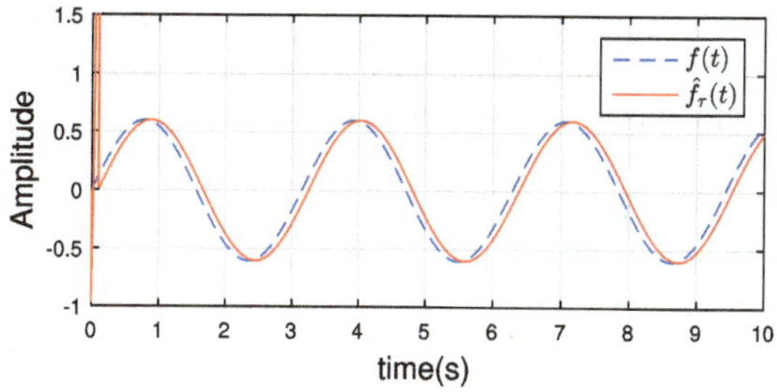

(a) Comparison between the fault signal and its estimate using the method presented in section 1.1.3 using the ISOH.

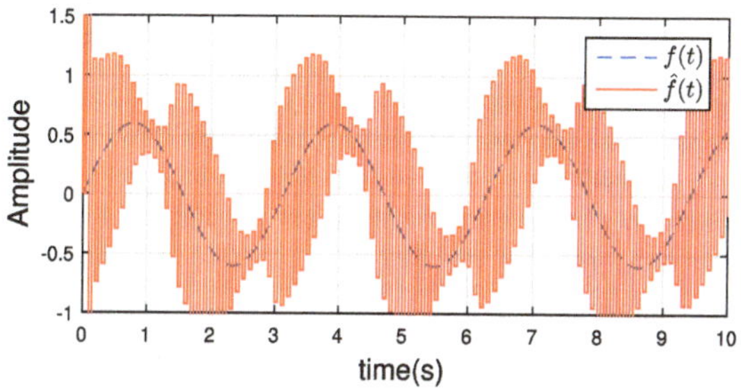

(b) Comparison between the fault signal and its estimate using the method presented in [17].

Fig. 1.6 Comparison of the reconstruction of a sinusoidal fault signal on a first-order system

1.3 Notes and References

The increasing industry demand for better reliability of equipment has led to the development of fault detection and isolation (FDI) techniques, allowing diagnostics and prognostics of the system health, aiming at safer operations [21]. Strategies for condition-based maintenance (CBM), whereby the equipment health is continuously monitored, can also increase its availability and thus reduce its downtime by avoiding long degraded-state operation or unnecessary preventive maintenance stops [22].

Sliding mode observers (SMO) have long been proposed for control systems [10, 36] (for more references about the early stages of SMO, see [31]). Its application for model-based fault reconstruction has gained growing attention since the seminal paper [12]. In the latter approach, the fault signal is reconstructed by manipulating the output error injection signal of the observer so that the amount of continuous control needed for the observer to remain on the sliding surface is achieved even in the presence of a fault. The injection signal can provide the fault signal reconstruction. Such method was further developed in [33, 34] where the observer gain optimization, based on the *bounded real lemma* [7], was proposed to improve the robustness with respect to model uncertainties. Also, a second-order sliding-mode observer for fault reconstruction was proposed in [27], using the super twisting algorithm for multivariable systems. Recently, a cascade of two SMO were applied to reconstruct faults in a class of noninfinitely observable descriptor systems [5, 6].

Systems with time delays were considered by several authors in the recent past [3, 14, 16, 24]. For a recent account of papers about this topic, see the recent book [26] and the bibliography therein.

To the best of our knowledge, in the literature on sliding-mode observers for time-delay systems, only a few publications are dedicated to the reconstruction of faults in systems with measurement delays. In [17], the delay is motivated by sampled-output systems, while in [18], more general output delay systems are considered. A common feature of the latter papers is that a delayed output error signal is employed to construct the observer. Unfortunately, this leads to undesirable chattering in the observer output injection term. The stability analysis relies on the well-known Lyapunov–Krasovskii's functionals that in turn lead to LMI conditions, appropriate from a numerical computation viewpoint. A singular perturbation approach was invoked to obtain the observer gains guaranteeing observation error system stability and reduction of the chattering effects. However, this appears to limit the allowable measurement delays or sampling periods. In contrast, arbitrarily large delays can be allowed using a cascade of high-gain observers as in [1, 9]. In these papers, the aim was to predict the current system state with application of the separation principle for control design. However, no unknown disturbance was allowed in their approaches as is the case of interest for fault reconstruction purposes. The same limitation applies to the partial differential equation (PDE) approach to develop current state observers of delay systems described in [24] so that its application for fault reconstruction is still an open problem.

In [19] and [13], adaptive observer structures for classes of nonlinear sampled-output systems presenting unknown constant parameters are designed. Although their method does not require output reconstruction filters like the zero-order hold, they use some knowledge of the disturbance dynamics to determine the unknown constant parameters.

In this chapter, one main feature was the formalization of the *time-shift approach*. It allows the design of SMO for arbitrarily long time-varying delays of the output measurement. Without invoking the Lyapunov–Krasovskii functionals, the SMO can be used for fault reconstruction provided the time-shifted system satisfies the usual minimum phase and relative degree one assumptions. As a further advantage of the proposed approach, uncertainties of the delay are allowed without compromising the realization of ideal sliding mode, thus avoiding undesirable chattering effects present in existing fault reconstruction methods. The price to be paid is that the reconstructed fault signal is also delayed by the corresponding time shift, which is acceptable for certain relevant applications, *e.g.*, for space or deep water systems when large communication delays are expected. For long delay durations, the delayed reconstruction can be applied in closed-loop control in cases where the system bandwidth is not so large or some prediction method to advance the fault reconstruction can be used [28]. Also, the ability to reconstruct the entire fault signal including its transient is interesting to identify short, intermittent faults.

Another highlight of this chapter was the application of the proposed method to sampled-output systems. By *deliberately* introducing a known constant delay, the output signal can be interpolated, and the fault can be reconstructed in continuous time. The possibility of using higher-order interpolations allows reconstructing faults in systems with longer sampling periods. In [13, 19] and [17], the maximum allowed sampling period is limited.

References

1. Ahmed-Ali T, Cherrier E, Lamnabhi-Lagarrigue F (2012) Cascade high gain predictors for a class of nonlinear systems. IEEE Trans Autom Control 57(1):224-229
2. Alwi H, Edwards C, Tan CP (2011) Fault detection and fault-tolerant control using sliding modes, 1st ed. Springer-Verlag, London. Advances in Industrial Control
3. Basin M, Rodriguez-Gonzalez J (2005) A closed-form optimal control for linear systems with equal state and input delays. Automatica 41(5):915–920
4. Burton JA, Zinober ASI (1986) Continuous approximation of variable structure control. Int J Syst Sci 17(6):875–885
5. Chan JCL, Tan CP, Trinh H, Kamal MAS (2019) State and fault estimation for a class of non-infinitely observable descriptor systems using two sliding mode observers in cascade. J Franklin Inst 356(5):3010–3029
6. Chan JCL, Tan CP, Trinh H, Kamal MAS, Chiew YS (2019) Robust fault reconstruction for a class of non-infinitely observable descriptor systems using two sliding mode observers in cascade. Appl Math Comput 350:78–92
7. Chilali M, Gahinet P (1996) $\mathscr{H}_\infty$ design with pole placement constraints: an LMI approach. IEEE Trans Autom Control 41(3):358–367
8. Cleveland WB (1976) First-order-hold Interpolation Digital-to-analog Converter with Application to Aircraft Simulation. NASA Technical Note D-8331
9. Coutinho CL, Oliveira TR, Cunha JPVS (2014) Output-feedback sliding mode control via cascade observers for global stabilisation of a class of nonlinear systems with output delay. Int J Control 87(11), 2327–2337
10. Edwards C, Spurgeon SK (1994) On the development of discontinuous observers. Int J Control 59(5):1211–1229
11. Edwards C, Spurgeon SK (1998) Sliding mode control: theory and applications. Taylor & Francis, London
12. Edwards C, Spurgeon SK, Patton RJ (2000) Sliding mode observers for fault detection and isolation. Automatica 36(4):541–553
13. Folin T, Ahmed-Ali T, Giri F, Burlion L, Lamnabhi-Lagarrigue F (2016) Sampled-data adaptive observer for a class of state-affine output-injection nonlinear systems. IEEE Trans Autom Control 61(2):462–467
14. Fridman E (2014) Introduction to time-delay systems: analysis and control, 1st edn. Birkhäuser, Basel. Systems & Control: Foundations & Applications
15. Fridman E (2014) Tutorial on Lyapunov-based methods for time-delay systems. Euro J Control 20(6):271–283
16. Gu K, Kharitonov VL, Chen J (2003) Stability of time-delay systems, 1st edn. Birkhäuser, Boston
17. Han X, Fridman E, Spurgeon SK (2014) Sampled-data sliding mode observer for robust fault reconstruction: a time-delay approach. J Franklin Inst 351(4):2125–2142

18. Han X, Fridman E, Spurgeon SK (2016) Sliding mode observer for robust fault reconstruction of time delay systems. In: Malisoff M, Pepe P, Mazenc F, Karafyllis I (eds) Recent results on nonlinear delay control systems: in honor of Miroslav Krstic. Springer International Publishing, London, pp 183–203
19. Hann CA, Ahmed-Ali T (2014) Continuous adaptive observer for state affine sampled-data systems. Int J Robust Nonlin Control 24(4):669–681
20. Hsu L, Costa RR, da Cunha JPVS (2003) Model-reference output-feedback sliding mode controller for a class of multivariable nonlinear systems. Asian J Control 5(4):543–556
21. Isermann R (2006) Fault-diagnosis systems: an introduction from fault detection to fault tolerance. Springer-Verlag, Berlin
22. Jardine A, Lin D, Banjevic D (2006) A review on machinery diagnostics and prognostics implementing condition-based maintenance. Mech Syst Sig Proces 20(7), 1483–1510
23. Khalil HK (2002) Nonlinear systems, 3rd edn. Prentice Hall, Saddle River
24. Krstic M (2009) Delay compensation for nonlinear, adaptive, and PDE systems. Birkhäuser, Basel
25. Liu Y, Park JH, Guo B-Z (2017) Results on stability of linear systems with time varying delay. IET Control Theory Appl 11(1):129–134
26. Malisoff M, Pepe P, Mazenc F, Karafyllis I (2016) Recent results on nonlinear delay control systems: in honor of Miroslav Krstic. Springer International Publishing, London. Series: Advances in Delays and Dynamics 4
27. Nagesh I, Edwards C (2014) A multivariable super-twisting sliding mode approach. Automatica 50:984–988
28. Pinto HLCP, Oliveira TR, Hsu L, Krstic M (2018) Sliding mode control for disturbance rejection in systems with measurement delay using PDE-backstepping predictor. The 2018 American Control Conference, pp 4099–4104
29. Seuret A, Gouaisbaut F (2013) Jensen's and Wirtinger's inequalities for time-delay systems. IFAC Proc Vol 46(3):343–348
30. Seuret A, Gouaisbaut F, Baudouin L (2016) D1.1 - overview of Lyapunov methods for time-delay systems. Research Report, Rapport LAAS n.16308, LAAS-CNRS
31. Spurgeon SK (2008) Sliding mode observers: a survey. Int J Syst Sci 39(8):751–764
32. Süli E, Mayers DF (2003) An introduction to numerical analysis. Cambridge University Press, Cambridge
33. Tan CP, Edwards C (2002) Sliding mode observers for detection and reconstruction of sensor faults. Automatica 38(10):1815–1821
34. Tan CP, Edwards C (2003) Sliding mode observers for robust detection and reconstruction of actuator and sensor faults. Int J Rob Nonlin Control 13(5):443–463
35. Utkin VI (1992) Sliding modes in control and optimization. Springer, Berlin
36. Walcott B, Zak S (1987) State observation of nonlinear uncertain dynamical systems. IEEE Trans Autom Control 32(2):166–170
37. Zhang Z, Chong KT (2007) Second order hold based discretization method of input time-delay systems. In: IEEE 2007 international symposium on information technology convergence (ISITC 2007), Jeonju, pp 348–352

Chapter 2
Fault Reconstruction for Delay Systems via Least-Squares and Time-Shifted SMOs

Building on the principles of delay-compensated estimation of Chap. 1, this chapter explores a quite general class of fault reconstruction in systems with delayed outputs using the same time-shifted sliding-mode observer (SMO). Here, the actuator faults are parametrized as a weighted sum of known regressor functions with unknown coefficients. Additionally, unlike the previous presentation in Chap. 1, where the state estimation and fault reconstruction were carried out only for the delayed signals, now a prediction scheme, leveraging the variation of constants formula, allows present-time estimation of unmeasured states. The fault is identified in real time using two continuous-time least-squares methods: the continuous-time recursive least-squares (RLS) algorithm with forgetting factor [22, section 4.3.6] for deterministic functions, and the continuous-time least-mean-squares (LMS) filter [27, 36] for stochastic functions. Theoretical results guarantee ideal sliding-mode operation, even amidst time-varying delays, preventing chattering in SMO-based output estimation. An application to petroleum engineering for failure diagnostics in an oil and gas intelligent completion valve demonstrates the method's effectiveness.

2.1 Problem Formulation

In this section, we present the problem formulation. The system assumptions as well as its representation in the time-shift approach are described. The disturbance model is presented as a sum of known regressor functions weighted by unknown constants [59].

Figure 2.1 presents a diagram for the proposed strategy.

H. L. C. P. Pinto et al., *Sliding-Mode Control and Estimation for Time-Delay Systems*, SpringerBriefs in Control, Automation and Robotics,
https://doi.org/10.1007/978-3-032-30178-9_2

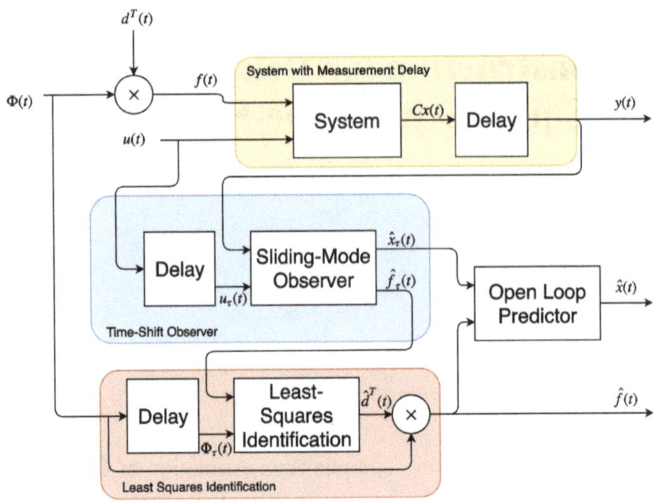

Fig. 2.1 Diagram for the proposed strategy

2.1.1 System Definition

Consider the following linear system presenting a known time-varying measurement delay $\tau(t) \geq 0$:

$$\begin{cases} \dot{x}(t) = Ax(t) + Bu(t) + Ff(t) \\ y(t) = Cx(t - \tau(t)) \end{cases}, \tag{2.1}$$

where $A \in \mathbb{R}^{n\times n}$, $B \in \mathbb{R}^{n\times m}$, $C \in \mathbb{R}^{p\times n}$, $F \in \mathbb{R}^{n\times l}$. The input signal $u(t) \in \mathbb{R}^m$ is known, the system state $x(t) \in \mathbb{R}^n$ is not measurable, and $y(t) \in \mathbb{R}^p$ is the output signal.

The unknown disturbance signal $f(t) \in \mathbb{R}^l$, $l \leq p < n$, represents the effect of an actuator fault. The following assumptions are made:

(A1) The matrices A, B, C, F are known. B and F have full rank.

(A2) The time delay $\tau(t)$ is differentiable and satisfies $0 < \tau(t) \leq \bar{\tau}$ for some constant $\bar{\tau}$, and its time derivative $\dot{\tau}(t)$ satisfies

$$\underline{r} \leq \dot{\tau}(t) \leq \bar{r} < 1, \quad |\dot{\tau}(t)| \leq r \tag{2.2}$$

where $r, \bar{r} > 0$, and $\underline{r}$ are constants, and $\underline{r}$ is possibly negative. These conditions must be simultaneously satisfied.

(A3) The unknown disturbance/fault $f(t)$ is norm bounded by

$$\|f(t)\| \leq \alpha, \quad \alpha \in \mathbb{R}^+. \tag{2.3}$$

(A4) The pair (A,C) is observable, and the invariant zeros of the triple (A, F, C) lie in the left half plane.
(A5) $rank(CF) = rank(F) = l$.
(A6) The past values of $u(t)$ for $t \in [t - \tau, t]$ are available for the fault reconstruction method.

Assumption **(A2)** is typically found in delay systems [14, 16, 31]. It establishes an upper bound for the delay and also guarantees that $t - \tau(t)$ is strictly increasing, meaning that the delayed variable affects the system following a chronological order [38, 47].

From **(A4)** and **(A5)**, it is assumed that the system (2.1) is already in the following canonical form suitable for observer design [11, Lemma 6.1]:

$$
\begin{aligned}
A &= \left[\begin{array}{c|c} A_{11} & A_{12} \\ \hline A_{21} & A_{22} \end{array}\right], \quad B = \left[\begin{array}{c} B_1 \\ \hline B_2 \end{array}\right], \\
F &= \left[\begin{array}{c} 0 \\ \hline F_2 \end{array}\right], \quad F_2 = \left[\begin{array}{c} 0 \\ \hline F_0 \end{array}\right] \\
C &= \left[\begin{array}{c|c} 0 & T \end{array}\right], \quad T^T T = I,
\end{aligned} \tag{2.4}
$$

where $A_{11} \in \mathbb{R}^{(n-p)\times(n-p)}$, $T \in \mathbb{R}^{p\times p}$ is orthogonal, and $B_2 \in \mathbb{R}^{p\times m}$, $F_2 \in \mathbb{R}^{p\times l}$. Matrix $F_0 \in \mathbb{R}^{l\times l}$ is nonsingular.

Remark 2.1 Assumptions **(A4)** and **(A5)** are only required for the system to be in this canonical form. The time-shift approach presented in this chapter can also be applied to different observer structures that do not require such assumptions.

The *objective* is to design an observer that can estimate both state and disturbance, at current time from the input signal $u(t)$ and the output signal $y(t)$ that presents an arbitrarily long time-varying delay $\tau(t)$.

2.1.2 Time-Shift Approach for Observer Design

Let us define the τ-delayed variables:

$$
\begin{aligned}
&x_\tau(t) = x(t - \tau(t)), \;\; f_\tau(t) = f(t - \tau(t)), \\
&u_\tau(t) = u(t - \tau(t)).
\end{aligned} \tag{2.5}
$$

Then, applying the chain rule for derivatives, system (2.1) can be described using the delayed state variables as follows:

$$
\begin{cases}
\dot{x}_\tau(t) = (1 - \dot{\tau}(t))\left[Ax_\tau(t) + Bu_\tau(t) + Ff_\tau(t)\right] \\
y(t) \;\;= Cx_\tau(t).
\end{cases} \tag{2.6}
$$

Notice that the time-shifted τ-delayed system has no (explicit) output delay. It is then possible to design an observer for the time-shifted system (2.6) as a generalization of the existing SMO, for systems presenting output delays. This observer is capable of reproducing the τ-delayed disturbance signal $f_\tau(t)$ and the τ-delayed state $x_\tau(t)$ from the output $y(t)$ and from the input $u(t)$.

2.1.3 Disturbance Model

In order to obtain a current estimate of the (nondelayed) disturbance signal $f(t)$ from its τ-delayed version $f_\tau(t)$, some knowledge of its nature is necessary. To this purpose, we consider that the disturbance signal $f(t) = [f_1 \quad f_2 \quad \cdots \quad f_l]^T$ has each of its $f_i(t)$ components modeled as

$$f_i(t) = d_{i0} + \sum_{k=1}^{\bar{k}} d_{ik}\phi_k(t), \quad i = 1, 2, \cdots, l, \tag{2.7}$$

with some unknown constants d_{i0} and d_{ik} and some known bounded continuous signals $\phi_k(t)$, $i = 1, 2, \cdots, l$, $k = 1, 2, \cdots, \bar{k}$. Defining:

$$\Phi(t) = \left[1 \; \phi_1(t) \; \phi_2(t) \; \cdots \; \phi_k(t)\right]^T, \tag{2.8}$$

$$d_i = \left[d_{i0} \; d_{i1} \; d_{i2} \; \cdots \; d_{ik}\right]^T, \tag{2.9}$$

each component $f_i(t)$ can also be written as

$$f_i(t) = d_i^T \Phi(t). \tag{2.10}$$

2.2 Time-Shifted Observer

In this section, an SMO is designed to reconstruct the delayed fault signal and estimate the delayed states from the time-shifted representation of the system. A convergence theorem for the SMO is also presented.

Based on assumption **(A6)**, the delayed input $u_\tau(t) = u(t - \tau(t))$ is available. Then, consider the following sliding-mode observer:

$$\begin{cases} \dot{\hat{x}}_\tau(t) = (1 - \dot{\tau}(t))\left[A\hat{x}_\tau(t) + Bu_\tau(t) - G_l e_y + G_n \nu\right] \\ \hat{y}(t) = C\hat{x}_\tau(t) \end{cases}, \tag{2.11}$$

where the state observation error $e(t)$ and the output observation error $e_y(t)$ are defined as

$$e(t) \triangleq \hat{x}_\tau(t) - x_\tau(t), \quad e_y(t) \triangleq \hat{y}(t) - y(t) = Ce(t). \tag{2.12}$$

From (2.6), (2.11), and (2.12), it can be seen that the observation error $e(t)$ is governed by

$$\dot{e}(t) = (1 - \dot{\tau}(t)) \left[(A - G_l C)e(t) + G_n \nu - F f_\tau(t)\right] \tag{2.13}$$

$$e_y(t) = Ce(t), \tag{2.14}$$

where G_l is chosen such that $A - G_l C$ is Hurwitz.

The discontinuous injection term ν is given by

$$\nu = -\rho \frac{P_0 e_y}{\|P_0 e_y\|}, \tag{2.15}$$

where $P_0 > 0$ is a design matrix. The modulation function or gain ρ is designed so that a sliding mode $e_y(t) \equiv 0$ is reached in finite time.

Remark 2.2 When the measurement delay $\tau(t)$ is constant ($\dot{\tau}(t) = 0$), equations (2.11) and (2.13) for the observer proposed here reduce to those of [12] originally proposed for fault reconstruction free of delays, but now estimating the delayed states rather than the current states.

Matrix G_n is chosen to have the structure:

$$G_n = \begin{bmatrix} -L \\ I_{p\times p} \end{bmatrix} T^T, \tag{2.16}$$

where T is orthogonal as defined in (2.4). Matrix $L = [L_1 \quad 0] \in \mathbb{R}^{(n-p)\times p}$, with $L_1 \in \mathbb{R}^{(n-p)\times(p-l)}$, is chosen such that $LF_2 = 0$ and $A_{11} + LA_{21}$ is Hurwitz, due to assumption **(A4)** (see [2, Lemma 4.3]), for $p < l$.

Remark 2.3 Note that if $p = l$, it will not be possible to determine matrix L. In this case, $\det(CF) \neq 0$ by assumptions **(A4)**–**(A5)**. Hence, the system (A, F, C) is minimum phase and relative degree one, and there exist $(n - p)$ invariant zeros, so that A_{11} is Hurwitz.

By a change of coordinates $e \mapsto \bar{e} = T_L e$, where T_L is a nonsingular linear transformation defined as

$$T_L = \begin{bmatrix} I_{n-p} & L \\ 0 & T \end{bmatrix}, \quad \bar{e} = \begin{bmatrix} \bar{e}_1 \\ \bar{e}_2 \end{bmatrix}, \quad e_y = \bar{e}_2, \tag{2.17}$$

the error dynamics can be described as

$$\begin{cases} \dot{\bar{e}}_1(t) = (1-\dot{\tau}(t))\left(\mathcal{A}_{11}\bar{e}_1(t) + (\mathcal{A}_{12} - \mathcal{G}_{l,1})e_y(t)\right) \\ \dot{e}_y(t) = (1-\dot{\tau}(t))\left(\mathcal{A}_{21}\bar{e}_1(t) + (\mathcal{A}_{22} - \mathcal{G}_{l,2})e_y(t)\right. \\ \qquad \left. +\nu - \mathcal{F}_2 f_\tau(t)\right) \end{cases}, \tag{2.18}$$

where matrices $(\mathcal{A}, C, \mathcal{F}, \mathcal{G}_l)$ correspond to the transformed matrices (A, C, F, G_l). Matrices $\mathcal{G}_{l,1}$ and $\mathcal{G}_{l,2}$ are partitions of the transformed matrix $\mathcal{G}_l = T_L G_l$ with the appropriate dimensions.

It can be seen that the sliding mode is governed by the eigenvalues of $\mathcal{A}_{11} = A_{11} + LA_{21}$, which is Hurwitz by design. Choosing $\mathcal{G}_{l,1} = \mathcal{A}_{21}$ and $(\mathcal{A}_{22} - \mathcal{G}_{l,2}) = \mathcal{A}_{22}^*$, where $\mathcal{A}_{22}^* \in \mathbb{R}^{p\times p}$ is any stable design matrix, the system (2.18) becomes triangular. Consequently, the eigenvalues of $A - G_l C$ will be those of $\mathcal{A}_{11}$ and $\mathcal{A}_{22}^*$. In the coordinate system of (2.4), G_l is then given by

$$G_l = T_L^{-1}\begin{bmatrix} \mathcal{A}_{12} \\ \mathcal{A}_{22} - \mathcal{A}_{22}^* \end{bmatrix}. \tag{2.19}$$

The symmetric design matrix $P_0 > 0$ of (2.15) must be selected satisfying $(\mathcal{A}_{22}^*)^T P_0 + P_0(\mathcal{A}_{22}^*) < 0$, which is always possible, given that $\mathcal{A}_{22}^*$ is Hurwitz by design. Also, there exists a matrix $P > 0$, satisfying $(A - G_l C)^T P + P(A - G_l C) < 0$, with the following structure [2]:

$$P = T_L^T \begin{bmatrix} P_1 & 0 \\ 0 & P_0 \end{bmatrix} T_L \tag{2.20}$$

The convergence of the observer (2.11) to the τ-delayed system (2.6) can be proved in the next theorem. This complete proof was not presented in [42].

Theorem 2.1 *Consider the τ-delayed system (2.6) that satisfies* **(A2)** *to* **(A6)** *and its observer (2.11), with (2.15), matrices (2.16) and (2.19), respectively. If the modulation function ρ of the discontinuous term (2.15) satisfies*

$$\rho \geq \alpha\|CF\| + \eta_0, \tag{2.21}$$

where η_0 is any positive scalar, then the state estimation error $e(t)$ converges exponentially to zero. The output error $e_y(t)$ reaches zero in finite time $t = t_s$, and a sliding mode starts on $\mathcal{S} = \{e : e_y = Ce = 0\}$, $\forall t \geq t_s$.

Proof As for Proposition 4.1 in [2] for systems without delay, consider the Lyapunov candidate $V_1(e) = e^T Pe$, with $e(t)$ from (2.13) and P from (2.20), for the state stability and convergence proof. Choose by design $PG_n \triangleq C^T P_0$ and $PF \triangleq C^T P_0 CF$. Differentiating the function along the trajectories of (2.13), applying the expression (2.15), and considering assumption **(A3)**, one has

$$\dot{V}_1(e) \leq (1-\dot{\tau}(t))\left(-\lambda_{\min}(Q)\|e\|^2 - 2\|P_0 e_y\|(\rho - \|CF\|\alpha)\right), \tag{2.22}$$

where $A_o^T P + P A_o = -Q$, $Q > 0$, $P > 0$. Under assumption **(A1)**, we have $(1-\dot{\tau}(t)) > 0$. Given that $-\lambda_{\min}(Q)\|e\|^2 < 0$, applying (2.21), we obtain

$$\dot{V}_1(e) \leq (1-\dot{\tau}(t))\left(-2\|P_0 e_y\|\eta_0\right) \tag{2.23}$$

$$\dot{V}_1(e) < 0, \quad \forall e \neq 0, \tag{2.24}$$

showing that the error $e(t)$ converges asymptotically.

For sliding-mode reachability, consider the Lyapunov candidate $V_2(e_y) = e_y^T P_0 e_y$, with $e_y(t)$ from (2.18) and P_0 from (2.20). Differentiating $V_2(e_y)$ along the trajectories of (2.18), considering that $(\mathcal{A}_{22}^*)^T P_0 + P_0(\mathcal{A}_{22}^*) < 0$, and applying (2.21), we can write the following inequality:

$$\dot{V}_2(\bar{e}_y) \leq (1-\dot{\tau})\left(2\|\bar{e}_y\|\left(\|P_0\mathcal{A}_{21}\bar{e}_1\| + \eta_0\right)\right). \tag{2.25}$$

From (2.24), it can be seen that the error e_1 reaches the set

$$\Omega_\eta = \{e_1 : \|P_0\mathcal{A}_{21}\bar{e}_1\| < \eta - \eta_0\}$$

in finite time and remains there, for a given scalar $\eta > 0$. The inequality inside this domain becomes

$$\dot{V}_2(\bar{e}_y) \leq (1-\dot{\tau})\left(-2\eta\|\bar{e}_y\|\right) \tag{2.26}$$

$$\dot{V}_2(\bar{e}_y) \leq -2\eta(1-\dot{\tau})\sqrt{\|P_0^{-1}\|}\sqrt{V_2}, \tag{2.27}$$

demonstrating that the output error e_y reaches zero in finite time and a sliding mode is initiated on the surface $\mathcal{S} = \{e : Ce = 0\}$. □

Once the sliding mode is reached in finite time, then $e_y(t) \equiv \dot{e}_y(t) \equiv 0$, and from (2.18), the error $\bar{e}_1(t)$ will be governed by

$$\begin{cases} \dot{\bar{e}}_1(t) = (1-\dot{\tau}(t))\left(\mathcal{A}_{11}\bar{e}_1(t)\right) \\ 0 \quad\; = (1-\dot{\tau}(t))\left(\mathcal{A}_{21}\bar{e}_1(t) + \nu_{eq} - \mathcal{F}_2 f_\tau(t)\right) \end{cases}, \tag{2.28}$$

where ν_{eq} is the equivalent output error injection signal (in the sense of equivalent control [53]), which represents the continuous control signal needed to maintain the sliding mode.

Corollary 2.1 *Under the conditions in Theorem 2.1, the delayed disturbance signal $f_\tau(t)$ in (2.5) can be reconstructed, with arbitrary accuracy, as $\delta \to 0$, by the* *expression*

$$\hat{f}_\tau(t) = \mathcal{F}_2^\dagger \nu_\delta, \tag{2.29}$$

where ν_δ *is the continuous approximation of* ν_{eq} *defined as*

$$\nu_\delta = -\rho \frac{P_0 e_y}{\|P_0 e_y\| + \delta}, \tag{2.30}$$

for a small scalar parameter $\delta > 0$*, and* $\mathcal{F}_2^\dagger$ *is the left pseudo-inverse of* $\mathcal{F}_2 = T_L F_2$*, given by*

$$\mathcal{F}_2^\dagger \triangleq (\mathcal{F}_2^T \mathcal{F}_2)^{-1} \mathcal{F}_2^T. \tag{2.31}$$

Proof Since, by assumption **(A5)**, the matrix F is full rank, the left pseudo-inverse for $\mathcal{F}_2$ exists as defined by (2.31). Therefore, the estimate $\hat{f}_\tau$ of the τ-delayed fault signal $f_\tau(t)$ is given by the expression (2.29). From Theorem 2.1, the sliding mode $e_y \equiv 0$ is reached in finite time, the state observation error is $\bar{e}_1(t) \to 0$ exponentially, and according to (2.28), one has

$$\nu_{eq} \to \mathcal{F}_2 f_\tau(t). \tag{2.32}$$

The equivalent output error injection ν_{eq} can be recovered by using the approximation ν_δ given by (2.30), with small enough choice of δ, to achieve any degree of accuracy [4]. □

The main advantage of the above time-shift approach is the possibility of determining the observer gains without appealing to specific methods for time-delay systems, like *Lyapunov–Krasovskii* or *Lyapunov–Razumikhin* approaches [13, 16]. In general, these methods restrict the maximum delay duration to some sufficiently small value. Also, these approaches usually result in linear matrix inequality (LMI) feasibility problems that add much analytical and numerical complexity to the observer design [47].

Note that when the time-shift approach is applied to design the observer, no delay is present in its feedback loop. As a result, the delay duration can be arbitrarily long, given that it is bounded. With the usual assumptions that the delay rate is strictly less than one and bounded from below [14, 31, 47], the problem is transformed to an observer for a time-varying system due to the term $(1 - \dot{\tau})$, and the Lyapunov function method can be applied in a straightforward manner. Although the state and faults are reconstructed with delay, this is a less conservative result than those obtained in [25, 26] in which convergence is limited to a certain delay range. In the proposed strategy, there is no limitation on this upper bound and no restriction on the delay range for the convergence.

In the present approach, an ideal sliding mode can theoretically be produced for arbitrarily long delays, which in practice also avoids low-frequency chattering problems. In [19], the observer is designed with delayed output, according to the original (nontime-shifted) system, and the discontinuous term depends on the delayed error signal. An ideal sliding mode is not achievable, and the reconstruction is subject to chattering effects (see [19, Remark 3]).

Remark 2.4 In the case of parametric or delay duration uncertainties, the time-shift procedure can still be used to design a sliding-mode observer similar to the one presented in [52]. In this approach, a constant upper bound can be used to model the uncertain delay, as in [26], while the delay and parametric uncertainties are lumped together in an unknown (yet bounded) term. This is a topic for a future publication.

2.3 Continuous-Time Least-Squares Identification

In this section, we present two methods to obtain an estimate of the weight vector $\hat{d}$, the continuous-time recursive least-squares (RLS) algorithm, suited for deterministic functions, and the continuous-time least-mean-squares (LMS) algorithms that can be applied to stochastic regressors.

2.3.1 *Recursive Least Squares*

Using the observer (2.11), an estimate of the delayed disturbance $\hat{f}_\tau(t)$ is obtained by Corollary 2.1. From (2.10), each component $\hat{f}_{\tau,i}(t)$ of $\hat{f}_\tau(t)$ can be described as

$$\hat{f}_{\tau,i}(t) = \hat{d}_i^T \Phi_\tau(t), \tag{2.33}$$

$$\hat{d}_i = \left[\hat{d}_{i0}\ \hat{d}_{i1}\ \hat{d}_{i2} \cdots \hat{d}_{ik} \right]^T, \tag{2.34}$$

where $\Phi_\tau(t) = \Phi(t - \tau(t))$, $\hat{d}_i$ is an estimate of d_i in (2.9), and the constants $\hat{d}_{ik}$ are estimates of its components d_{ik}, $i = 1, 2, \cdots, l$, $k = 1, 2, \cdots, \bar{k}$. Since the delayed estimate of the disturbance signal $\hat{f}_\tau(t)$ and the delayed generating functions $\Phi_\tau(t)$ are available, it is possible to obtain an estimate of the weight vector $\hat{d}$ by means of an identification method.

Least squares is one of the oldest and most used identification methods. It fits a mathematical model in the form presented in (2.10) to a sequence of observed data, in our case $\hat{f}_\tau(t)$, minimizing the sum of the squares of the difference between the observed and computed data. In our particular case, we will be using the continuous-time recursive least-squares (RLS) algorithm with forgetting factor [22, section 4.3.6], for each component $\hat{f}_{\tau,i}(t)$ of $\hat{f}_\tau(t)$.

Consider the normalized error:

$$e_{\hat{f}_{\tau,i}} = \frac{\hat{f}_{\tau,i} - \hat{d}_i(t)^T \Phi_\tau(t)}{\mu^2}, \tag{2.35}$$

$$\mu^2 = 1 + \Phi_\tau(t)^T \Phi_\tau(t), \tag{2.36}$$

where $\hat{d}_i(t)$ is the instantaneous value for the estimate of $\hat{d}_i$ in the RLS algorithm and μ^2 is the normalizing factor. This factor is not critical in this case since it is known that $\hat{f}_{\tau,i}$ and $\Phi_\tau(t)$ are bounded.

The algorithm determines a *covariance matrix* K_i, using the differential equation:

$$\dot{K}_i = \beta K_i - K_i \frac{\Phi_\tau(t)\Phi_\tau(t)^T}{\mu^2} K_i, \quad K_i(0) = K_{i,0}, \tag{2.37}$$

where the initial estimate $K_{i,0}$ is a symmetric positive-definite matrix. The forgetting factor $\beta > 0$ prevents that K_i becomes too small, slowing or stalling the adaptation process. On the other hand, choosing a too large value for β can lead K_i to become unstable. In this case, it is possible to use a norm bound for the covariance matrix setting $\dot{K}_i = 0$ when $\|K_i\| > R_0$, where $R_0 > 0$ is a design constant matrix.

The estimates $\hat{d}_i$ are updated using

$$\dot{\hat{d}}_i(t) = K_i e_{\hat{f}_{\tau,i}} \Phi_\tau(t). \tag{2.38}$$

This algorithm is equivalent to minimizing the cost function:

$$\begin{aligned} J(\hat{d}_i) = &\frac{1}{2}\int_0^t e^{-\beta(t-\theta)} \frac{\left(\hat{f}_{\tau,i} - \hat{d}_i^T \Phi_\tau(t)\right)^2}{\mu^2} d\theta \\ &+ \frac{1}{2} e^{-\beta t} \left(\hat{d}_i(t) - \hat{d}_i(0)\right)^T K_{i,0}^{-1} \left(\hat{d}_i(t) - \hat{d}_i(0)\right). \end{aligned} \tag{2.39}$$

From the cost function, it can be seen that the forgetting factor β minimizes the influence of both the initial choice $\hat{d}_i(0)$ and past measured values. Convergence of the algorithm can be adjusted by selecting an appropriate $\hat{d}_i(0)$ and by adjusting β.

One of the main advantages of the adaptive law (2.37)–(2.38) is that it can be shown [22, Theorem 4.3.2] that $e_{\hat{f}_{\tau,i}}$, $\dot{\hat{d}}_i(t)$, and $\hat{d}_i(t)$ are bound even if $\Phi_\tau(t)$ is not. Also, it guarantees that if $\Phi_\tau(t)$ is bound and persistently exciting the estimate, $\hat{d}_i(t)$ converges exponentially to the optimum value $\hat{d}_i$.

Using the estimate for the weight vector $\hat{d}_i(t)$, an estimate of the current fault $\hat{f}(t)$ can be obtained from the generating functions $\Phi(t)$ using the expression:

$$\hat{f}_i(t) = \hat{d}_i(t)^T \Phi(t), \quad i = 0, 1, \cdots, l, \tag{2.40}$$

$$\hat{f}(t) = \left[\hat{f}_1(t) \quad \hat{f}_2(t) \quad \cdots \quad \hat{f}_q(t)\right]^T. \tag{2.41}$$

Notice that since the regressors are known and the coefficients are already estimated, the fault reconstruction in (2.40) is performed by several vector inner products or by a matrix to vector multiplication.

Let us define the disturbance estimation error:

$$e_f(t) \triangleq \hat{f}(t) - f(t). \tag{2.42}$$

Corollary 2.1 shows that the delayed estimate $\hat{f}_\tau(t) \to f_\tau(t)$ as $t \to \infty$. Using the estimates $\hat{d}_i$, it can be seen that

$$e_f(t) \triangleq \left(\hat{d}_i(t) - d_i\right)^T \Phi(t). \tag{2.43}$$

Since $\hat{d}_i(t)$ is obtained from the estimate of the delayed disturbance $\hat{f}_\tau(t)$, $e_f(t)$ will be minimized with $e_{\hat{f}_{\tau,i}}$ by the identification algorithm, but it will still be dependent from the precision of the reconstruction given by the choice of δ in (2.30).

2.3.2 Continuous-Time Least-Mean-Squares Filter

The RLS algorithm presented in the previous section and in [42] can only be applied when the regressor functions are deterministic. When the regressors are stochastic functions, one option for identification is to apply a continuous-time least-mean-squares (LMS) filter [27, 36] described here. This is one of the most widely used algorithm in adaptive filtering for its low computational complexity and convergence in stationary environment [9, Chap. 3]. Indeed, consider the statistical average of the squared error:

$$\begin{aligned} \xi_i(t) &= \mathbf{E}[(\hat{f}_{\tau,i}(t) - \hat{d}_i^T \Phi_\tau(t))^2] \\ &= \mathbf{E}[\hat{f}_{\tau,i}(t)^2] - 2\mathcal{P}_i^T \hat{d}_i + \hat{d}_i^T \mathcal{R}_i \hat{d}_i, \end{aligned} \tag{2.44}$$

where $\mathbf{E}[\cdot]$ is the statistical mean of $[\cdot]$,

$$\mathcal{P}_i = \mathbf{E}[\hat{f}_{\tau,i}(t)\Phi_\tau(t)], \mathcal{R}_i = \mathbf{E}[\Phi_\tau(t)\Phi_\tau^T(t)], \tag{2.45}$$

and $\hat{f}_{\tau,i}(t)$ and $\hat{d}_i^T$ are as defined in (2.33) and (2.34).

Vector $\mathcal{P}_i \in \mathbb{R}^{(\bar{k}+1)}$ is the cross correlation between $\hat{f}_{\tau,i}(t)$ and $\Phi_\tau(t)$ at zero lag, and $\mathcal{R}_i \in \mathbb{R}^{(\bar{k}+1)\times(\bar{k}+1)}$ is the autocorrelation matrix of $\Phi_\tau(t)$ at zero lag. One important property of $\mathcal{R}_i$ is that it is positive semidefinite. Since in practice the regressors are usually linear independent, it can be assumed as being positive definite [9, Section 2.3]. Also, since $\mathcal{R}_i$ is a Hermitian matrix, all its eigenvalues are real valued.

The minimum of ξ_i can be found by setting the gradient with respect to $\hat{d}_i$ to zero:

$$\nabla_{\hat{d}_i}\xi_i = -2\mathcal{P}_i + 2\mathcal{R}_i\hat{d}_i = 0. \tag{2.46}$$

The Wiener optimal solution is then $\hat{d}_{i,\text{opt}} = \mathcal{R}_i^{-1}\mathcal{P}_i$. This can also be obtained by finding the solution when $t \to \infty$ of

$$\frac{d}{dt}\hat{d}_i(t) = -\gamma\mathcal{R}_i\hat{d}_i(t) + \gamma\mathcal{P}_i, \tag{2.47}$$

where $\gamma > 0$ is any positive constant. The autocorrelation matrix $\mathcal{R}_i$ and the cross-correlation vector $\mathcal{P}_i$ are not readily available, so we propose the following continuous-time estimate:

$$\mathcal{P}_i(t) = \begin{cases} \hat{f}_{\tau,i}(0)\Phi_\tau(0) & , t = 0 \\ \frac{1}{t}\int_0^t \hat{f}_{\tau,i}(\theta)\Phi_\tau(\theta)d\theta & , t > 0 \end{cases} \tag{2.48}$$

$$\mathcal{R}_i(t) = \begin{cases} \Phi_\tau(0)\Phi_\tau^T(0) & , t = 0 \\ \frac{1}{t}\int_0^t \Phi_\tau(\theta)\Phi_\tau^T(\theta)d\theta & , t > 0 \end{cases}. \tag{2.49}$$

Equation (2.47) becomes

$$\frac{d}{dt}\hat{d}_i(t) = -\gamma\mathcal{R}_i(t)\hat{d}_i(t) + \gamma\mathcal{P}_i(t), \tag{2.50}$$

which is the continuous-time LMS algorithm. Equation (2.50) is a linear time-varying, forced system. From assumption **(A3)** and taking into account all elements of $\Phi(t)$ are continuous and bounded, it can be seen that $\mathcal{P}_i(t)$ is also bounded. Therefore, it can be verified that the system (2.50) is BIBO (bounded input, bounded output) with respect to $\gamma\mathcal{P}_i(t)$ and that the norm of $\hat{d}_i(t)$ is bounded by [27, eq (17)]:

$$\|\hat{d}_i(t)\| \leq \|\hat{d}_i(0)\|e^{-\gamma\mu_R t} + \gamma\int_0^t e^{-\gamma\mu_R(t-\theta)}\|P_i(\theta)\|d\theta, \tag{2.51}$$

where $0 < \mu_R < \lambda_{\min}(\mathcal{R}_i(t))$, with $\lambda_{\min}(\mathcal{R}_i(t))$ being the minimum eigenvalue of $\mathcal{R}_i(t)$. From (2.51), it can be seen that the speed of convergence depends on the parameter γ and the initial condition $\hat{d}_i(0)$.

After the convergence, an estimate of the current fault $\hat{f}(t)$ can be obtained using (2.40) as with the RLS algorithm.

2.4 Predictor Based on the Variation of Constants Formula

This section presents a predictor based on the variation of constants formula using the signals obtained from Sects. 2.2 and 2.3.

The system (2.1) is linear; hence, the current state $x(t)$ can be obtained from the delayed state $x_\tau(t)$, using the variation of constants formula [61]:

$$x(t) = e^{A\tau(t)} x_\tau(t) + \int_{t-\tau(t)}^{t} e^{A(t-\theta)} \left[Bu(\theta) + Ff(\theta)\right] d\theta \tag{2.52}$$

if all signals were known.

Using the estimates $\hat{x}_\tau(t)$ and $\hat{f}_\tau(t)$ from the observer (2.11), an estimate of the disturbance at current time $\hat{f}(t)$ is obtained using one of the identification algorithms from the previous section. Then, substituting $x_\tau(t)$ for $\hat{x}_\tau(t)$ and $f(t)$ for $\hat{f}(t)$, we can obtain an estimate of the current state:

$$\hat{x}(t) = e^{A\tau(t)} \hat{x}_\tau(t) + \int_{t-\tau(t)}^{t} e^{A(t-\theta)} \left[Bu(\theta) + F\hat{f}(\theta)\right] d\theta. \tag{2.53}$$

The prediction error $e_x(t) = \hat{x}(t) - x(t)$ will then be given by

$$e_x(t) = e^{A\tau(t)} e(t) + F \int_{t-\tau(t)}^{t} e^{A(t-\theta)} e_f(\theta) d\theta. \tag{2.54}$$

In Theorem 2.1, it is shown that the state estimation error $e(t)$ converges exponentially to zero. Hence, it can be seen that the estimation error will reach its minimum as $e_f(t)$ does. Also, since the error is integrated during a period of length $\tau(t)$, this residual will also be smaller for shorter delays.

Table 2.1 presents the complete equations for the proposed observer-predictor strategy illustrated in Fig. 2.1.

2.5 Petroleum Engineering Application

In our previous conference publication [42], only an academic example was used to demonstrate the proposed strategy. Here, we use a petroleum engineering application to demonstrate the use of the proposed strategy method in a real-world scenario.

Table 2.1 Summary of the proposed method

(1) System	i. System description: $\begin{cases} \dot{x}(t) = Ax(t) + Bu(t) + Ff(t) \\ y(t) = Cx(t-\tau(t)) \end{cases}$ ii. Fault model: $\Phi(t) = \left[1\ \phi_1(t)\ \phi_2(t) \cdots \phi_k(t)\right]^T$, $d_i = \left[d_{i0}\ d_{i1}\ d_{i2} \cdots d_{ik}\right]^T$, $f_i(t) = d_i^T\Phi(t)$.
(2) Time-shifted system representation	i. Delayed variables: $x_\tau(t) = x(t-\tau(t)),\ f_\tau(t) = f(t-\tau(t)),$ $u_\tau(t) = u(t-\tau(t))$. ii. Time-shifted representation: $\begin{cases} \dot{x}_\tau(t) = (1-\dot{\tau}(t))\left[Ax_\tau(t) + Bu_\tau(t) + Ff_\tau(t)\right] \\ y(t) = Cx_\tau(t) \end{cases}$
(3) Time-shifted SMO	$\begin{cases} \dot{\hat{x}}_\tau(t) = (1-\dot{\tau}(t))\left[A\hat{x}_\tau(t) + Bu_\tau(t) - G_l e_y + G_n \nu\right] \\ \hat{y}(t) = C\hat{x}_\tau(t) \end{cases}$
(4) (a) RLS fault identification (deterministic regressors)	i. Normalized error: $e_{\hat{f}_\tau,i} = \frac{\hat{f}_{\tau,i} - \hat{d}_i(t)^T\Phi_\tau(t)}{\mu^2}$, $\mu^2 = 1 + \Phi_\tau(t)^T\Phi_\tau(t)$. ii. Covariance matrix: $\dot{K}_i = \beta K_i - K_i \frac{\Phi_\tau(t)\Phi_\tau(t)^T}{\mu^2} K_i$, $K_i(0) = K_{i,0}$. iii. Estimate update: $\dot{\hat{d}}_i(t) = K_i e_{\hat{f}_\tau,i}\Phi_\tau(t)$.
(4) (b) LMS fault identification (stochastic regressors)	i. Cross-correlation estimate: $\mathcal{P}_i(0) = \hat{f}_{\tau,i}(0)\Phi_\tau(0)$, $\mathcal{P}_i(t) = \frac{1}{t}\int_0^t \hat{f}_{\tau,i}(\theta)\Phi_\tau(\theta)d\theta$, $\forall t > 0$ ii. Autocorrelation estimate: $\mathcal{R}_i(0) = \Phi_\tau(0)\Phi_\tau^T(0)$, $\mathcal{R}_i(t) = \frac{1}{t}\int_0^t \Phi_\tau(\theta)\Phi_\tau^T(\theta)d\theta$, $\forall t > 0$ iii. Estimate update: $\frac{d}{dt}\hat{d}_i(t) = -\gamma\mathcal{R}_i(t)\hat{d}_i(t) + \gamma\mathcal{P}_i(t)$.
(5) Fault prediction	$\hat{f}_i(t) = \hat{d}_i(t)^T\Phi(t)$, $i = 0, 1, \cdots, l$.
(6) Open-loop state prediction	$\hat{x}(t) = e^{A\tau(t)}\hat{x}_\tau(t) + \int_{t-\tau(t)}^{t} e^{A(t-\theta)}\left[Bu(\theta) + F\hat{f}(\theta)\right]d\theta$

Inside an oil well, a set of pipes named *production string* is installed to conduct the oil or gas to surface. In this production string, several equipment are installed to allow the control of production. Among them is the intelligent completion valve (ICV) [41, 45], which provides control of the production for a given reservoir segment (interval). Such valves have a very high cost of acquisition and installation. In consequence, they are supposed to have a very long lifetime (15 years or more), justifying the efforts to implement CBM and FDI strategies. The simple dynamics of the valves and the presence of a large measurement delay motivate the adoption of the proposed strategy in order to detect and, if possible, compensate faults, postponing a costly replacement intervention.

One very common failure mode in the ICVs is inorganic salt scaling of the sleeve [15, 17], which impairs production by clogging the valve orifices. Scaling also increases the friction force on the sleeve, which affects the actuation mechanism, sometimes stalling it. The idea is to use the proposed structure as a diagnostics tool for ICVs, allowing to estimate variations in the friction coefficient and determining velocity and position behavior.

Remark 2.5 This chapter is focused on the actuator fault reconstruction, assuming that the sensors are working properly. A complete CBM implementation should also include sensor fault detection schemes [24, 29, 51, 60], avoiding that such faults would be misinterpreted as actuation problems.

2.5.1 Simplified Model of an ICV

The ICV is comprised of a sliding sleeve valve actuated by a dual action hydraulic piston that allows for remote operation from surface. These valves are typically installed in depths greater than 1000 ft below sea floor. Only the pressure applied to the opening line and the fluid flow rates (from both opening and return lines) are available for monitoring. Figure 2.2 illustrates a simplified schematic of its actuation mechanism.

In real-case scenarios, rigid capillary hydraulic lines are used and the control fluids are degassed. Then, the valve model will be simplified by assuming that the fluid is incompressible, and there is no elasticity of the materials. In this case, the fluid supply flow rate (q_s), the piston input flow rate (q_i), and output flow rate (q_o) can be considered equal by conservation of mass ($q_s = q_i = q_o$). Also, an ideal pressure supply is being considered, such that the supply pressure (p_s) does not depend on the supply flow rate.

Since the control line diameter is much smaller than its length and because the actuation is typically slow, it can be considered that the flow is always laminar and the line has a linear resistive characteristic, which is described by Watton [58] and Merritt [37]

$$\Delta p_{pipe} = r_{pipe} \cdot q_{pipe} \tag{2.55}$$

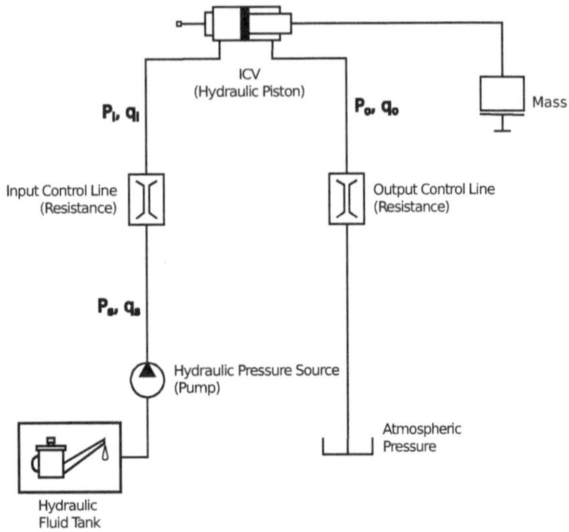

Fig. 2.2 Hydraulic schematic of the simplified model of an intelligent completion valve actuation mechanism. The fluid is pumped through the input control line to the opening chamber of the piston. As it moves, hydraulic fluid is displaced and pushed through the output (return) control line to the atmospheric tank. Both control lines offer some resistance to the passage of fluid

$$r_{pipe} = \frac{128\mu_{fluid}\ell_{pipe}}{\pi d_{pipe}^4}, \tag{2.56}$$

where μ_{fluid} is the fluid viscosity in centistokes, ℓ_{pipe} and d_{pipe} are the length and diameter of the pipe in meters, Δp_{pipe} is the pressure differential on the pipe, r_{pipe} is its hydraulic resistance, and q_{pipe} is the flow rate.

Since the piston is symmetrical, only one direction of movement will be analyzed. Let m_p be the total displaced mass of the piston (valve sleeve and its own mass), x_p its position, S_p the piston area, and $F_r(\dot{x}_p)$ the total friction force applied to the piston. Using Newton's second law of motion, one obtains

$$m_p\ddot{x}_p = S_p(p_i - p_o) - F_r(\dot{x}_p), \tag{2.57}$$

where p_i, p_o are the pressure in the input and output chambers of the piston, respectively. Using the assumption of fluid incompressibility and considering the linear behavior of the hydraulic lines, p_i and p_o can be described as [37, 58]

$$(p_s - p_i) = r_i q_i \implies p_i = p_s - r_i q_i \tag{2.58}$$

$$(p_o - 0) = r_o q_o \implies p_o = r_o q_o, \tag{2.59}$$

where r_i, r_o are the input and output total flow resistances, respectively. By conservation of mass, $q_s = q_i = q_o = S_p\dot{x}_p$, and from (2.58) and (2.59):

$$p_i = p_s - r_i S_p\dot{x}_p, \quad p_o = r_i S_p\dot{x}_p. \tag{2.60}$$

Then, applying (2.60) in (2.57), we obtain

$$m_p \ddot{x}_p = S_p p_s - S_p^2 (r_i + r_o) \dot{x}_p - F_r(\dot{x}_p). \tag{2.61}$$

The output flow rate is proportional to the velocity of the piston ($\dot{x}_p$), and it will be assumed as the system output. Since it is a fluid displacement, the propagation occurs at the speed of sound in the hydraulic fluid [33] being modeled here by an output delay, given by

$$\tau_o = \ell_o v_s, \tag{2.62}$$

where ℓ_o is the length of the hydraulic line and v_s is the speed of sound in the fluid. For our simplified model, since the pipeline is modeled as only a restriction (resistive characteristic) with a constant delay, we can manipulate the supply pressure p_s to completely compensate for any input delay, so this will not be represented in the model.

Defining the state variables $x_1 = x_p$, $x_2 = \dot{x}_p$, and the input signal $u = p_s$, it is possible to write the state equations:

$$\begin{cases} \dot{x} = \underbrace{\begin{bmatrix} 0 & 1 \\ 0 & -\frac{S_p^2(r_i+r_o)}{m_p} \end{bmatrix}}_{A} x + \underbrace{\begin{bmatrix} 0 \\ \frac{S_p}{m_p} \end{bmatrix}}_{B} u + \underbrace{\begin{bmatrix} 0 \\ \frac{F_r(x_2)}{m_p} \end{bmatrix}}_{Ff(x_2(t))} \\ y = \underbrace{\begin{bmatrix} 0 & S_p \end{bmatrix}}_{C} x(t - \tau_o). \end{cases} \tag{2.63}$$

It is also necessary to model the hard stop at the end of the piston displacement. The simplest form to implement it in a computer simulation is as a saturation:

$$x_{sat} = \begin{cases} \begin{bmatrix} 0 \\ 0 \end{bmatrix}, & x_1 < 0, x_2 < 0; \\ \begin{bmatrix} x_{p,\max} \\ 0 \end{bmatrix}, & x_1 > x_{p,\max}, x_2 > 0; \\ x, & \text{other cases}, \end{cases} \tag{2.64}$$

where $x_{p,\max}$ is the maximum piston displacement and x_{sat} is the state including the saturations.

There are several models available for the friction force [28, sec 1.2.3]. Since a typical valve sleeve moves very slowly, the *Static plus Coulomb plus Linear viscous friction* combination was chosen:

$$\begin{cases} F_r(\dot{x}_p) = F_S, & \dot{x}_p < v_{th} \\ F_r(\dot{x}_p) = F_C + k_1\dot{x}_p, & \dot{x}_p \geq v_{th}, \end{cases} \tag{2.65}$$

where F_S is the static friction force, F_C is the Coulomb friction force near to zero velocity, k_1 is the friction coefficient, and v_{th} is the threshold speed, where the change from static to viscous friction occurs. From this expression, it can be seen that after the threshold, the friction can be modeled as in (2.7), with $d_0 = F_C$, $d_1 = k_1$, and $\phi_1(t) = \dot{x}_p$.

2.5.2 Numerical Simulations for Valve Diagnostics

Due to the nature of the friction, it can be seen that the unknown signal $f(x_2(t))$ is bounded, satisfying assumption **(A3)**. Hence, the time-shifted observer can be used to obtain a delayed estimate of the friction force. Since matrix C in (2.63) is known (the piston area is known), the delayed variable $x_2(t - \tau_o)$ is measurable by dividing the output flow rate (q_o, measured) by the piston area (S_p). In addition, if the velocity of the sleeve is varied so that x_2 becomes a persistently exciting signal, the RLS algorithm is guaranteed to converge [22, Theorem 4.3.4], allowing the estimates for F_C and k_1 in (2.65) to be obtained, which is the desired diagnostic information regarding the friction force. Finally, applying an implicit version of (2.53),

$$\hat{x}(t) = e^{A\tau_o}\hat{x}_\tau + \int_{t-\tau_o}^{t} e^{A(t-\theta)}\left[Bu(\theta) + F\hat{f}(x_2(\theta))\right]d\theta, \tag{2.66}$$

we obtain present time estimates of the state variables, completing the diagnostics.

A numerical simulation was performed in order to validate this strategy. Table 2.2 shows the ICV parameters utilized for the simulation. This valve is designed to move with a differential actuation pressure greater than 3000 psi.

Due to the very large values of some parameters, the velocity equation was scaled down by a factor of 10^{-6}, avoiding the matrix A becoming bad conditioned. Applying the parameter values, the system matrices for the state representation in the regular form (2.4) are

$$A = \begin{bmatrix} 0 & -0.001 \\ 0 & -1.96 \end{bmatrix}, \quad B = \begin{bmatrix} 0 \\ 10^{-8} \end{bmatrix}, \quad F = \begin{bmatrix} 0 \\ 10^{-5} \end{bmatrix},$$
$$C = \begin{bmatrix} 0 & 1 \end{bmatrix}. \tag{2.67}$$

Notice that when the state variables were used in other calculations (such as friction force or the hard stop), they were scaled up by a factor of 10^6 so that the friction and the hard-stop functions could be still described in SI units. The actuation pressure

Table 2.2 Simulation parameters for the ICV

Valve parameter	Value (in SI units)
Mass (m_p)	100 Kg
Piston area (S_p)	10^{-6} m^2
Pipe length (ℓ_{pipe})	600 m
Total input resistance (r_i)	1.47×10^{14} Pa · s/m^3
Total output resistance (r_o)	4.91×10^{13} Pa · s/m^3
Delay (τ_o)	0.4107 s
Static friction (F_S)	2.08×10^4 N[a]
Coulomb friction (F_C)	2.08×10^4 N[b]
Friction coefficient (k_1)	2.60×10^4 N · s/m
Max. piston displacement ($x_{p,\max}$)	0.3 m

[a] Equivalent to 3000 psi pressure on the piston
[b] Equivalent to 2900 psi pressure on the piston

$P_s = 3010 + 100\sin(15t)$ (psi) was designed to shift the valve and generate a persistently exciting velocity.

The piston position (x_1) is nonobservable, but the initial condition for both state variables is assumed known, since the piston can be moved to the hard stop at position $x_1 = 0$, $x_2 = 0$ before diagnostics. Note that, even if in this simplified model we assume that the delayed state $x_2(t - \tau_o)$ is measurable, this might not be true in other applications. Hence, a time-shifted observer for the piston velocity (x_2) can be designed to reconstruct the delayed friction force $f(x_2(t - \tau_o))$.

Since the system is already in the regular form (2.4), the fault reconstruction time-shifted observer (2.11) is designed as follows:

1. Choose the stable design matrix $\mathcal{A}_{22}^* \in \mathbb{R}^{(p\times p)}$ and compute matrix G_l in (2.19) using the partitions of $\mathcal{A}$. This solution is nonunique. We have set $\mathcal{A}_{22}^* = [-10]$ to allocate the poles of $A - G_l C$ in $\{0, -10\}$, taking into account that the pole in 0 is from an nonobservable state and therefore cannot be changed. In theory, this could result in an unbounded signal in the closed-loop system. To avoid this problem, the saturation function (hard stop) in (2.64) is implemented in the observer to keep x_1 bounded.
2. From (2.16) and (2.19), we obtain

$$G_n = \begin{bmatrix} 0 \\ 1 \end{bmatrix}, \quad G_l = \begin{bmatrix} 0.001 \\ 0.008 \end{bmatrix}. \tag{2.68}$$

3. Choose P_0 satisfying $(\mathcal{A}_{22}^*)^T P_0 + P_0(\mathcal{A}_{22}^*) < 0$. In our case, $P_0 = 0.05$ was used.
4. Determine ρ satisfying (2.21). In our case, a constant $\rho = 2.7 \times 10^4$ was used.
5. To reconstruct the fault, it was used $\delta = 0.5 \times 10^{-6}$ for the approximation of the equivalent control v_δ in (1.22).

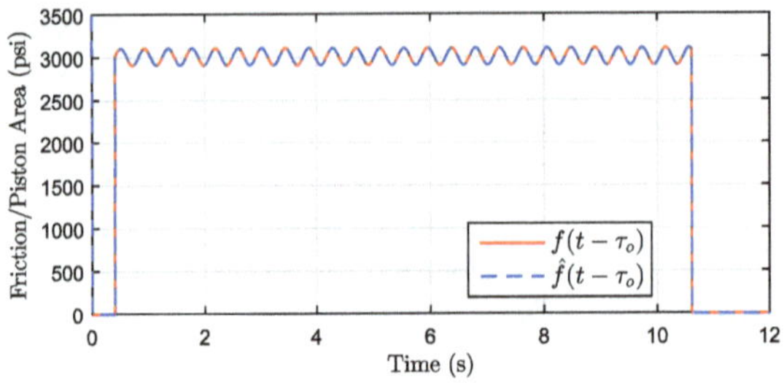

Fig. 2.3 Comparison of real and estimated delayed reconstruction of friction $F_r(x_2(t-\tau_o))$ obtained using the time-shifted observer from Corollary 2.1. The valve was actuated with a pressure $P_s = 3010 + 100\sin(15t)$. The friction is represented as force per piston area in psi to allow comparison to the actuation pressure

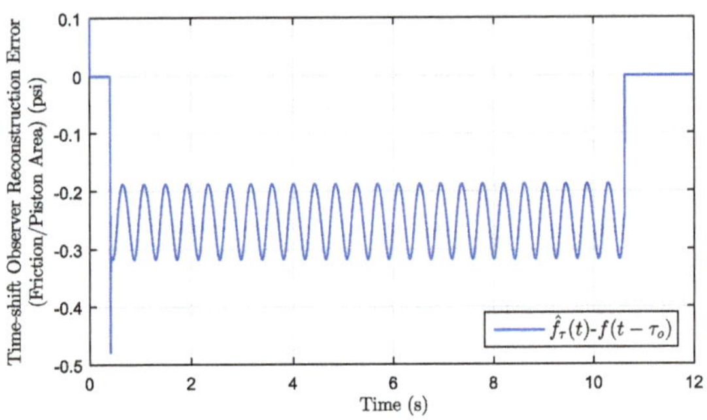

Fig. 2.4 Estimation error for the signals presented in Fig. 2.3. The maximum residual error is of the order of 0.5 psi, which is very small when compared to the amplitudes of the friction signal $f(t)$ (around 3000 psi)

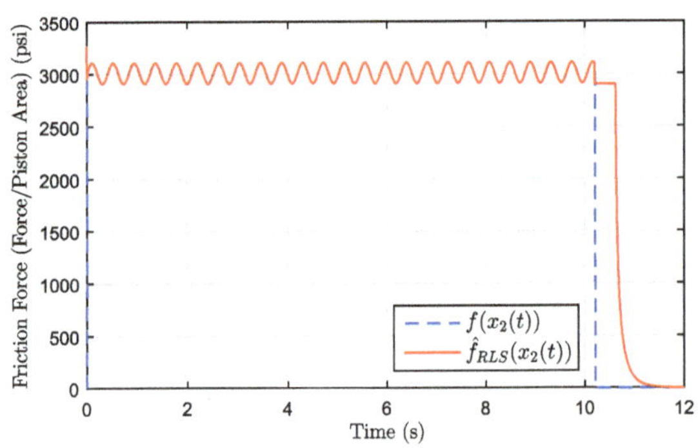

Fig. 2.5 Comparison of real and estimated values of the friction using the RLS identification algorithm with $\beta = 5$

Figures 2.3 and 2.4 compare the real value of $f(x_2(t-\tau_o))$ with the estimated value $\hat{f}(x_2(t-\tau_o))$, showing a very accurate reconstruction (less than 1 psi difference). The friction is represented as force per piston area in psi to allow comparison to the actuation pressure.

Utilizing the estimate of the friction force $\hat{f}(x_2(t-\tau_o))$, it is possible to use the RLS algorithm in Eqs. (2.37) and (2.38) with the measured delayed state $x_2(t-\tau_o)$ as the known base function $\phi_1(t-\tau_o)$ from (2.33) to estimate the parameters $d_0 = F_C$ and $d_1 = k_1$ of (2.65). Figure 2.5 compares the real and estimated

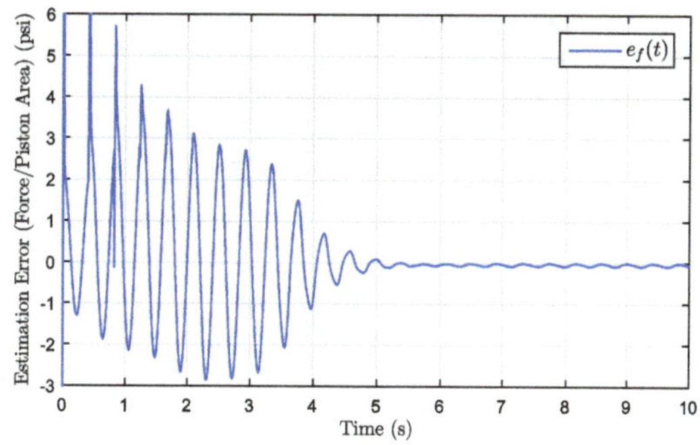

Fig. 2.6 Estimation error of the friction force in (2.43) for the signals presented in Fig. 2.5

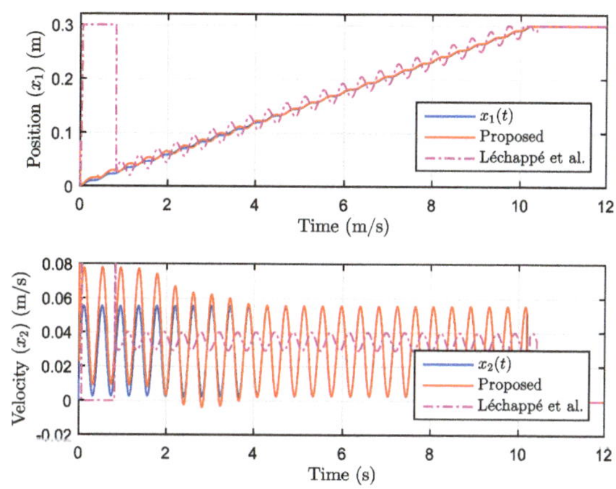

Fig. 2.7 State variables x_1, x_2 compared to their respective estimates $\hat{x}_1, \hat{x}_2$, obtained using the proposed method (Table 2.1), and their estimates obtained using the predictor from [34]

values of the friction using the RLS identification algorithm with the forgetting factor $\beta = 5$ in (2.37). Figure 2.6 shows the estimation error of the friction force $e_f(t) = f(x_2(t)) - \hat{f}_{RLS}(x_2(t))$, where $\hat{f}_{RLS}(x_2(t))$ is the RLS identified estimate of $f(x_2(t))$. After some transient, the error rapidly decays to less than 2 psi. The convergence time can be reduced if the initial choices for the weights are closer to the real values, which would happen, in consecutive actuations of the valve and more rounds of algorithm iterations.

The performance of the proposed observer-predictor strategy was compared to the one presented in [34] where the delayed estimate for the fault is used as the estimate for the current time fault in Eq. (2.66). Figures 2.7 and 2.8 compare the state variables with their estimates obtained using the open-loop predictor in (2.66), showing good convergence of the proposed strategy after a few seconds. The position and velocity estimates may be used to verify if the valve has shifted correctly. On the other hand, since the disturbance is fast varying, the strategy proposed by Léchappé et al. in [34] estimates the states with larger residual errors.

The method applied here considered that the delayed state $x_2(t - \tau_o)$ is measurable, which might not be true in other applications. In fact, if the compressibility of the fluid was considered in the valve model, it would be necessary to take into account the dynamics of the capillary lines, and the output would not be the velocity.

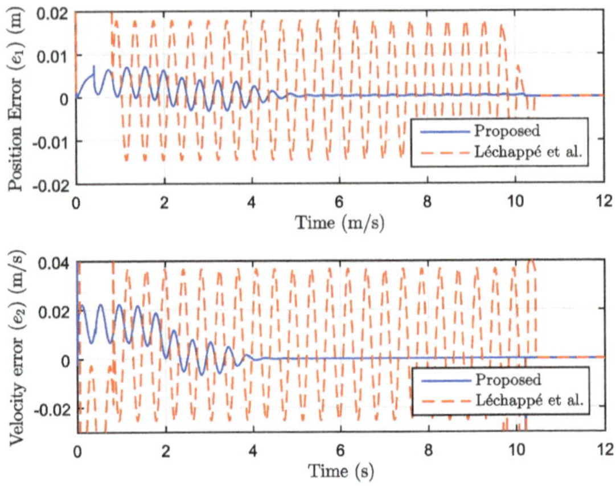

Fig. 2.8 Estimation error for the signals presented in Fig. 2.7

In this case, the estimate $\hat{x}_2(t - \tau_o)$ obtained from the time-shifted observer could be used in the RLS algorithm and in the open-loop predictor.

2.6 Notes and References

As discussed at the end of Chap. 1, strategies for condition-based maintenance (CBM), whereby the equipment health is continuously monitored, can increase system availability and thus reduce its downtime by avoiding long degraded-state operation or unnecessary preventive maintenance stops [23, 24]. In this context, several fault detection and isolation (FDI) techniques have been proposed [21], allowing diagnostics and prognostics of the system health. In systems designed for oil and gas wells, these techniques are still in their initial studies of application, with very few preliminary results [48].

For systems with parametric uncertainties or unmeasured disturbances, sliding-mode control (SMC) is a well-known and attractive control strategy, due to its robustness, good transient performance, and disturbance rejection capability [53], demonstrated by several applications and design advances [20, 40, 43], including control of fuzzy systems [55, 56], synchronization of chaotic systems [5, 6], and robust control of electrohydraulic actuators [60].

A feedforward compensator can be designed to attenuate or eliminate the effects of a disturbance or fault, if they can be measured or estimated by an observer [39]. A comprehensive overview of the disturbance observer designs can be seen in the survey papers [44] and [7]. Sliding-mode observers (SMO) have long been proposed for control systems [10, 54] (for more references about the early stages of SMO, see [50]). Their application for model-based fault reconstruction has gained growing attention since the publication of the paper [12], where the fault signal is reconstructed by manipulating the error injection signal of the observer. Some more recent publications show excellent results for FDI and fault-tolerant control applications [2, 29].

On the other hand, time delays are present either on actuation or sensing paths in most control engineering applications [30]. The presence of measurement delays in control loops may cause poor performance or instability. Systems with time delays were considered by several authors in the recent past [31, 35]. To the best of our knowledge, in the literature on SMOs for time-delay systems, only a few publications are dedicated to the reconstruction of faults in systems with measurement delays. In [18] and [19], a delayed output error signal is employed to construct the observer, which leads to undesirable chattering in the output injection term. Here, the time-shift sliding-mode observer, proposed in Chap. 1, can theoretically achieve an ideal sliding mode, with no chattering even in the presence of arbitrarily long delays.

Open-loop predictors can be used to overcome delays with known duration [3, 49, 57]. For linear systems, they can be derived from the analytical solution of the differential equations using the variation of constants formula. In [61], an open-loop predictor is used to advance the state and design an adaptive control law in order to stabilize a plant presenting an input delay. An open-loop predictor is also used with a delayed state observer in [32] and [31, Sec. 6.3] to obtain state estimates at current time in systems with measurement delays with no unknown inputs or disturbances. In [46], an approximation of the disturbance is obtained by a truncated Taylor series approximation. For longer delays, this method requires higher-order derivatives of the predicted signals, making it more sensitive to numerical errors.

This chapter developed an observer-predictor strategy that can estimate both state and disturbance, at current time from the input and the output signal, which presents an arbitrarily long time-varying delay. The objective of this chapter was to present a method to reconstruct faults and estimate state in current time of systems presenting time-varying output delays of arbitrary duration, based on SMO. Unlike other approaches in the literature [1, 8, 19], we have assumed unknown disturbances (faults) and time-varying output delays of arbitrary duration.

References

1. Ahmed-Ali T, Cherrier E, Lamnabhi-Lagarrigue F (2012) Cascade high gain predictors for a class of nonlinear systems. IEEE Trans Autom Control 57(1):224–229
2. Alwi H, Edwards C, Tan CP (2011) Fault detection and fault-tolerant control using sliding modes, 1st edn. Springer-Verlag, London. Advances in Industrial Control
3. Artstein Z (1982) Linear systems with delayed controls: a reduction. IEEE Trans Autom Control **27**(4):869–879
4. Burton JA, Zinober ASI (1986) Continuous approximation of variable structure control. Int J Syst Sci 17(6):875–885
5. Chen X, Park JH, Cao J, Qiu J (2017) Sliding mode synchronization of multiple chaotic systems with uncertainties and disturbances. Appl Math Comput 308:161–173
6. Chen X, Park JH, Cao J, Qiu J (2018) Adaptive synchronization of multiple uncertain coupled chaotic systems via sliding mode control. Neurocomputing 2017:9–21

7. Chen WH, Yang J, Guo L, Li S (2016) Disturbance-observer-based control and related methods: an overview. IEEE Ttans Ind Electron 63(2):1083–1095
8. Coutinho CL, Oliveira TR, Cunha JPVS (2014) Output-feedback sliding mode control via cascade observers for global stabilisation of a class of nonlinear systems with output delay. Int J Control 87(11):2327–2337
9. Diniz PSR (2002) Adaptive filtering: algorithms and practical implementation. Springer, Berlin
10. Edwards C, Spurgeon SK (1994) On the development of discontinuous observers. Int J Control 59(5):1211–1229
11. Edwards C, Spurgeon SK (1998) Sliding mode control: theory and applications. Taylor & Francis, London
12. Edwards C, Spurgeon SK, Patton R (2000) Sliding mode observers for fault detection and isolation. Automatica 36(4):541–553
13. Fridman E (2014) Introduction to time-delay systems: analysis and control, 1st ed,. Birkhäuser, Basel. Systems & Control: Foundations & Applications
14. Fridman E (2014) Tutorial on Lyapunov-based methods for time-delay systems. Euro J Control 20(6):271–283
15. Graham GM, Bezerra MCM, Goodwin N, Albino EH, Pinto HLCP, Bhavsar RB (2013) Minimizing scale deposition through surface enhancement in downhole tools. In: OTC Brasil
16. Gu K, Kharitonov VL, Chen J (2003) Stability of time-delay systems, 1st ed. Birkhäuser, Boston
17. Hamid S, De Jesus O, Jacinto CMC, Izetti RG, Pinto HLCP, Drougett EL, Edwards C, Cassidy J, Zhang H, Dagenais P, Batocchio M (2013) A practical method of predicting chemical scale formation in well completions. In: SPE Asia Pacific Oil and Gas Conference Exhibition
18. Han X, Fridman E, Spurgeon SK (2014) Sampled-data sliding mode observer for robust fault reconstruction: a time-delay approach. J Franklin Inst 351(4):2125–2142
19. Han X, Fridman E, Spurgeon SK (2016) Sliding mode observer for robust fault reconstruction of time delay systems. In: Malisoff M, Pepe P, Mazenc F, Karafyllis I (eds.) Recent results on nonlinear delay control systems: In honor of Miroslav Krstic. Springer International Publishing, London, pp 183–203
20. Hsu L, Costa RR, da Cunha JPVS (2003) Model-reference output-feedback sliding mode controller for a class of multivariable nonlinear systems. Asian J Control 5(4):543–556
21. Hwang I, Kim S, Kim Y, Seah CE (2010) A survey of fault detection, isolation, and reconfiguration methods. IEEE Trans Control Syst Technol 18(3):636–653
22. Ioannou PA, Sun J (1996) Robust adaptive control (out of print). Prentice Hall, Hoboken. Available online at: http://wwwrcf.usc.edu/~ioannou/Robust_Adaptive_Control.htm
23. Jardine A, Lin D, Banjevic D (2006) A review on machinery diagnostics and prognostics implementing condition-based maintenance. Mech Syst Sig Proces 20(7):1483–1510
24. Kandukuri ST, Klausen A, Karimi HR, Robbersmyr KG (2016) A review of diagnostics and prognostics of low-speed machinery towards wind turbine farm-level health management. Renew Sustainab Energy Rev 53(1):697–708
25. Karimi HR (2012) A sliding mode approach to H-infinity synchronization of master-slave time-delay systems with Markovian jumping parameters and nonlinear uncertainties. J Franklin Inst 349(4):1480–1496
26. Karimi HR, Gao H (2010) New delay-dependent exponential H-infinity synchronization for uncertain neural networks with mixed time delays. IEEE Trans Syst Man Cybernet Part B: Cybernet 40(1):173–185
27. Karni S, Zeng G (1989) The analysis of the continuous-time LMS algorithm. IEEE Trans Acoust Speech 37(4):595–597
28. Khalil HK (2002) Nonlinear systems, 3rd edn. Prentice Hall, Upper Saddle River
29. Kommuri SK, Defoort M, Karimi HR, Veluvolu KC (2016) A robust observer-based sensor fault-tolerant control for PMSM in electric vehicles. IEEE Trans Ind Electron 63(12):7671–7681
30. Krstic M (2010) Compensation of infinite-dimensional actuator and sensor dynamics. IEEE Contr Syst Mag 30(1):22–41

31. Krstic M (2009) Delay compensation for nonlinear, adaptive, and PDE systems. Birkhäuser, Basel
32. Krstic M, Smyshlyaev A (2008) Backstepping boundary control for first-order hyperbolic PDEs and application to systems with actuator and sensor delays. Syst Control Lett 57(9):750–758
33. Krus P, Weddfelt K, Palmberg J (1994) Fast pipeline models for simulation of hydraulic systems. J Dyn Syst-T ASME 116(1):132–136
34. Léchappé V, Moulay E, Plestan F, Glumineau A, Chriette A (2015) New predictive scheme for the control of LTI systems with input delay and unknown disturbances. Automatica 52:179–184
35. Malisoff M, Pepe P, Mazenc F, Karafyllis I (2016) Recent results on nonlinear delay control systems: in honor of Miroslav Krstic. Springer International Publishing, London. Series: Advances in Delays and Dynamics 4
36. Mandic DP, Kanna S, Constantinides AG (2015) On the intrinsic relationship between the least mean square and Kalman Filters [Lecture Notes]. IEEE Signal Proc Mag 32(6):117–122
37. Merritt HE (1991) Hydraulic control systems. Wiley, Hoboken
38. Michiels W, Verriest EI (2011) A look at fast varying and state dependent delays from a systems theory point of view. TW Reports, Department of Computer Science, K.U.Leuven; Leuven
39. Mohammadi A, Marquez HJ, Tavakoli M (2017) Nonlinear disturbance observers: design and applications to Euler-lagrange systems. IEEE Contr Syst Mag 37(4):50–72
40. Oliveira TR, Leite AC, Peixoto AJ, Hsu L (2014) Overcoming limitations of uncalibrated robotics visual servoing by means of sliding mode control and switching monitoring scheme. Asian J Control 16(3):752–764
41. Pinto HLCP, Izetti RG, Silva Jr MF, Guimaraes GB (2006) Integrated multi-zone low cost intelligent completion for mature fields. In: SPE intelligent energy conference and exhibition
42. Pinto HLCP, Oliveira TR, Hsu L (2018) Time-shift sliding mode predictor for fault reconstruction of time-delay systems using least squares identification. 15th international workshop on variable structure systems and sliding mode control (VSS18)
43. Queiroz K, Araujo A, Dias S (2012) Design and stability analysis of a variable structure adaptive backstepping controller. Asian J Control 14(3):641–651
44. Radke A, Gao Z (2006) A survey of state and disturbance observers for practitioners. In: IEEE 2006 American Control Conference - Minneapolis, MN, pp 5183–5188
45. Rahman J, Allen C, Bhat G (2012) Second generation interval control valve (ICV) improves operational efficiency and inflow performance in intelligent completions. In: IADC/SPE Asia Pacific Drilling Technology Conference & Exhibition
46. Sanz R, Garcia P, Albertos P (2016) Enhanced disturbance rejection for a predictor-based control of LTI systems with input delay. Automatica 72:205–208
47. Seuret A, Gouaisbaut F, Baudouin L (2016) D1.1 - Overview of Lyapunov methods for time-delay systems. Research Report, Rapport LAAS n.16308. LAAS-CNRS
48. Silva Jr MF, Radespiel ES, Jacinto CMC (2013) A diagnostic and prognostic framework for integrated reservoir-completion management using intelligent well data. In: 2013 offshore technology conference OTC Brasil
49. Smith OJM (1957) Closer control of loops with dead time. Chem Eng Progr 53(5):217–219
50. Spurgeon SK (2008) Sliding mode observers: a survey. Int J Syst Sci 39(8):751–764
51. Tan CP, Edwards C (2002) Sliding mode observers for detection and reconstruction of sensor faults. Automatica 38(10):1815–1821
52. Tan CP, Edwards C (2003) Sliding mode observers for robust detection and reconstruction of actuator and sensor faults. Int J Robust Nonlin Control 13(5):443–463
53. Utkin VI (1992) Sliding modes in control and optimization. Springer, Berlin
54. Walcott B, Zak S (1987) State observation of nonlinear uncertain dynamical systems. IEEE Trans Autom Control 32(2):166–170
55. Wang Y, Gao Y, Karimi HR, Shen H, Fang Z (2018) Sliding mode control of fuzzy singularly perturbed systems with application to electric circuit. IEEE Trans Syst Man Cy A PP(99):1–9

56. Wang Y, Shen H, Karimi HR, Duan D (2018) Dissipativity-based fuzzy integral sliding mode control of continuous-time T-S fuzzy systems. IEEE Trans Fuzzy Syst 26(3):1164–1176
57. Watanabe K, Ito M (1981) An observer for linear feedback control laws of multivariable systems with multiple delays in controls and outputs. Syst Control Lett 1(1):54–59
58. Watton J (2007) Modelling, monitoring and diagnostic techniques for fluid power systems, 1st edn. Springer, London
59. Wen L, Tao G, Liu Y (2016) Multivariable adaptive output rejection of unmatched input disturbances. Int J Adapt Control 30(8–10):1203–1227
60. Zhang H, Liu X, Wang J, Karimi HR (2014) Robust H_∞ sliding mode control with pole placement for a fluid power electrohydraulic actuator (EHA) system. Int J Adv Manuf Tech 73(5–8):1095–1104
61. Zhou B, Liu Q, Mazenc F (2017) Stabilization of linear systems with both input and state delays by observer-predictors. Automatica 83:368–377

Part II
Disturbance Rejection

Chapter 3
Disturbance Rejection and Estimation with PDE-Backstepping Predictor

Even though the proposed estimation scheme of Chap. 2 could be applied to closed-loop control following the steps shown in [28, Section IV], the controller design was not in the scope of Part I, but it is the precise pursuit of Part II of this brief.

In particular, in this chapter, we address the disturbance rejection problem for linear plants with delayed measured output. The output delay is allowed to be time-varying and of arbitrary duration. The disturbances that may represent actuator faults are generated by an exogenous dynamic system and can be matched or unmatched. A predictor-based observer is designed via the PDE-backstepping (partial differential equation-backstepping) method so that an estimate of the current state vector, as well as that of unmeasured disturbance, can be obtained. These estimates are used to regulate the output and reject the disturbance by output feedback sliding-mode control. In addition, a solution for finite-time disturbance rejection (rather than exponential only) is also introduced for the special case when the sensor delay is constant.

3.1 Problem Formulation

3.1.1 System Definition

Consider a linear time-invariant system, presenting an unknown, unmeasurable disturbance signal $d(t)$, which can be matched or unmatched, and a known time-varying measurement delay $\tau(t)$:

$$\begin{cases} \dot{x}(t) = Ax(t) + Bu(t) + Fd(t) \\ y(t) = Cx(t - \tau(t)), \end{cases} \tag{3.1}$$

H. L. C. P. Pinto et al., *Sliding-Mode Control and Estimation for Time-Delay Systems*, SpringerBriefs in Control, Automation and Robotics,
https://doi.org/10.1007/978-3-032-30178-9_3

where matrices $A \in \mathbb{R}^{n\times n}$, $B \in \mathbb{R}^{n\times 1}$, $C \in \mathbb{R}^{1\times n}$, $F \in \mathbb{R}^{n\times 1}$ are known. In this chapter, both input $u(t) \in \mathbb{R}$ and output signals $y(t) \in \mathbb{R}$ are assumed available; however, the system state $x(t) \in \mathbb{R}^n$ and the disturbance signal $d(t) \in \mathbb{R}$ are not directly measurable. The objective is to design for such system an output feedback control law that rejects the output effects of a persistent disturbance regulating its output to zero [45].

The following assumptions are made:

(A1) The known time-varying delay $\tau(t)$ is assumed differentiable, being positive and uniformly bounded from above such that: $0 \le \tau(t) \le \bar{\tau}$.

(A2) The time derivative of the delay $\dot{\tau}(t)$ exists, is sufficiently small, and is norm bounded (strictly smaller than one), ensuring that $t - \tau(t)$ is monotonically nondecreasing:

$$|\dot{\tau}(t)| \le h, \quad 0 \le h < 1 . \tag{3.2}$$

(A3) B is full rank, i.e., $\mathrm{rank}(B) = 1$.

(A4) The pair (A, C) is observable.

(A5) The pair (A, B) is controllable.

In particular, the finite upper bounds $\bar{\tau}$ in **(A1)** and h **(A2)** are also known since the time-varying delay $\tau(t)$ is assumed to be known in **(A1)** as well.

By means of a linear transformation, the system can be taken to the following form:

$$x = \begin{bmatrix} x_1 \\ x_2 \end{bmatrix}, \quad A = \begin{bmatrix} A_{11} & A_{12} \\ A_{21} & A_{22} \end{bmatrix}, \quad B = \begin{bmatrix} 0 \\ B_2 \end{bmatrix}, \quad F = \begin{bmatrix} F_1 \\ F_2 \end{bmatrix},$$
$$C = \begin{bmatrix} C_1 & C_2 \end{bmatrix}, \tag{3.3}$$

where $x_1 \in \mathbb{R}^{n-1}$, $x_2 \in \mathbb{R}$, $A_{11} \in \mathbb{R}^{(n-1)\times(n-1)}$, $C_1 \in \mathbb{R}^{1\times(n-1)}$, $C_2 \in \mathbb{R}$, and $B_2 \in \mathbb{R}$ is not null. Notice that $F_2 \in \mathbb{R}$ represents the matched portion of F, while $F_1 \in \mathbb{R}^{n-1}$ represents the unmatched one.

In the form of (3.3), a further assumption is also made:

(A6) Matrix A_{11} is Hurwitz, *i.e.*, the system is minimum phase from $u(t)$ to $x_2(t)$.

3.1.2 Structured Disturbance

In industrial plants, it is common practice to perform a reliability study during the design phase to anticipate failure modes based on prior information regarding their nature [17]. For this chapter, it is assumed that $d(t)$ can be modeled as the output of a linear dynamic exosystem [3, 29]:

$$\begin{cases} \dot{x}_d(t) = A_d x_d(t) \\ d(t) \;\; = C_d x_d(t) \end{cases}, \tag{3.4}$$

where $A_d \in \mathbb{R}^{n_d \times n_d}$ and $C_d \in \mathbb{R}^{1 \times n_d}$, and it is assumed that neither $d(t)$ nor the state $x_d(t) \in \mathbb{R}^{n_d}$ are directly measurable.

It is also assumed that:

(A7) The pair (A_d, C_d) is observable.

This is a basic condition for the design of the disturbance observer.

Regarding matrix A_d, to ensure that the disturbance signal is bounded, it is assumed that:

(A8) No eigenvalue of A_d has positive real part, and the eigenvalues with zero real parts are simple.

Even though matrices (A_d, C_d) are assumed known, disturbance $d(t)$ and its state vector x_d are to be estimated.

3.1.3 Augmented System

In order to develop an unknown input observer [29], an augmented representation combining the system (3.1) and the disturbance model (3.4) is considered. With state vector $z(t) \triangleq \left[x_d^T(t) \quad x^T(t)\right]^T$, the augmented system can be written as

$$\begin{cases} \dot{z}(t) = A_a z(t) + B_a u(t) \\ y(t) = C_a z(t - \tau(t)) \end{cases}, \tag{3.5}$$

where

$$A_a \triangleq \left[\begin{array}{c|c} A_d & 0 \\ \hline FC_d & A \end{array}\right], \quad B_a \triangleq \left[\begin{array}{c} 0 \\ \hline B \end{array}\right], \quad C_a \triangleq \left[\begin{array}{c|c} 0 & C \end{array}\right]. \tag{3.6}$$

Assumptions **(A4)** and **(A7)** guarantee that the augmented system (A_a, C_a) is also observable, reminding that the complete state vector is defined by $z := [x_d^T \;\; x_1^T \;\; x_2]^T$ (in its expanded form) and matrices A and B are the same which we have in (3.1), or in the corresponding form (3.3).

3.2 PDE-Backstepping-Based Unknown Input Observer

An unknown input observer will be derived for the augmented system with output delay, based on the PDE-backstepping [22, 36], using a transport partial differential equation (PDE) to model the time-varying measurement delay. Thereby, it is possible to predict the current state even in the presence of the disturbance.

3.2.1 Observer Equations

The measurement delay of the augmented system (3.5) can be represented by a transport partial differential equation (PDE) as

$$\begin{cases} \dot{z}(t) &= A_a z(t) + B_a u(t) \\ v_t(\ell, t) &= \pi(\ell, t) v_\ell(\ell, t) \\ v(1, t) &= C_a z(t) \\ y(t) &= v(0, t) \quad (= C_a z(t - \tau(t))) \end{cases}, \tag{3.7}$$

where the speed of propagation, given by

$$\pi(\ell, t) = \left(\dot{\tau}(t)(\ell - 1) + 1\right)/\tau(t), \tag{3.8}$$

is chosen such that the solution for this transport PDE is

$$v(\ell, t) = C_a z(t + (\ell - 1)\tau(t)). \tag{3.9}$$

The observer developed in what follows is an extension of the one presented in [22, Section 3.1] for constant measurement delays, described by

$$\begin{cases} \dot{\hat{z}}(t) &= A_a \hat{z}(t) + B_a u(t) + e^{A_a \tau(t)} L\left(y(t) - \hat{v}(0, t)\right) \\ \hat{v}_t(\ell, t) &= \pi(\ell, t)\hat{v}_\ell(\ell, t) + C_a e^{A_a \ell \tau(t)} L\left(y(t) - \hat{v}(0, t)\right), \\ \hat{v}(1, t) &= C_a \hat{z}(t) \end{cases} \tag{3.10}$$

where $\hat{z}(t) = [\hat{x}_d^T(t) \quad \hat{x}^T(t)]^T$ is the estimate of the augmented state z, composed by the state estimates of the disturbance, $\hat{x}_d$, and of the system, $\hat{x}$. Matrix L is chosen such that $A_a - LC_a$ is Hurwitz. The spatial variable $\ell \in [0, 1]$ is mapped to the measurement delay in the domain $[\tau(t), 0]$ so that $\ell = 1$ corresponds to zero delay and $\ell = 0$ corresponds to $\tau(t)$.

3.2.2 Convergence Analysis of the Proposed Observer

The *PDE-backstepping* method allows a formal proof of Lyapunov stability for the observation error system. Indeed, introducing the error variables:

$$\tilde{z}(t) = z(t) - \hat{z}(t), \quad \tilde{v}(\ell, t) = v(\ell, t) - \hat{v}(\ell, t), \tag{3.11}$$

the observer error dynamics is then given by

$$\begin{cases} \dot{\tilde{z}}(t) &= A_a\tilde{z}(t) - e^{A_a\tau(t)}L\tilde{v}(0,t) \\ \tilde{v}_t(\ell,t) &= \pi(\ell,t)\tilde{v}_\ell(\ell,t) - C_a e^{A_a\ell\tau(t)}L\tilde{v}(0,t) \\ \tilde{v}(1,t) &= C_a\tilde{z}(t) \end{cases}. \tag{3.12}$$

Based on the *PDE-backstepping* proposed for observers for systems presenting constant measurement delay [22, Section 3.1], [24], and using the variable ℓ as in the observer (3.10), the following transformation is obtained:

$$\tilde{w}(\ell,t) = \tilde{v}(\ell,t) - C_a e^{A_a(\ell-1)\tau(t)}\tilde{z}(t), \tag{3.13}$$

in order to bring the transport PDE in (3.12) to an undisturbed (homogeneous) form:

$$\tilde{w}_t(\ell,t) = \pi(\ell,t)\tilde{w}_\ell(\ell,t), \tag{3.14}$$

$$\tilde{w}(1,t) = 0. \tag{3.15}$$

It can be seen from the boundary condition (3.15) that the transformed variable $\tilde{w}(\ell, t)$ converges in finite time to zero.

Applying this transformation to (3.12), the observer error dynamics can be written as

$$\begin{cases} \dot{\tilde{z}}(t) &= \left(A_a - e^{A_a\tau(t)}LC_a e^{-A_a\tau(t)}\right)\tilde{z}(t) - e^{A_a\tau(t)}L\tilde{w}(0,t) \\ \tilde{w}_t(\ell,t) &= \pi(\ell,t)\tilde{w}_\ell(\ell,t) \\ \tilde{w}(1,t) &= 0 \end{cases}, \tag{3.16}$$

which has a convenient triangular structure.

Using a Lyapunov–Krasovskii functional [22], it can be demonstrated that the observer state (3.10) converges to the system state (3.5). A formal proof for the case of time-varying delay follows hereafter.

Theorem 3.1 *The observer error system (3.12) for the proposed observer (3.10) is exponentially stable at the origin in the sense of the norm:*

$$\left(\|z(t)-\hat{z}(t)\|^2+\int_0^1 \left(v(\ell,t)-\hat{v}(\ell,t)\right)^2 d\ell\right)^{1/2}, \tag{3.17}$$

and consequently, that $(\hat{z}, \hat{v})$ *exponentially converges to* (z, v).

Proof Based on assumptions **(A1)** and **(A2)** made on the delay and on its derivative, it can be observed that

$$\pi(0,t)=\frac{1-\dot{\tau}(t)}{\tau(t)}\geq \pi_0>0. \tag{3.18}$$

Let $\bar{M}$ be an upper bound for $\|e^{A_a\tau(t)}\|$, where $\|e^{A_a\tau(t)}\| \leq e^{\|A_a\|\bar{\tau}} < \bar{M}$. It is important to notice that since $A_a - LC_a$ is Hurwitz by design, the matrix $A_o \triangleq \left(A_a - e^{A_a\tau(t)}LC_a e^{-A_a\tau(t)}\right)$ will present stable eigenvalues for any value of $\tau(t)$. If $\tau(t)$ changes slowly enough, it can be shown [19, Lemma 9.9] that the Lyapunov equation $P(\tau(t))A_o + A_o^T P(\tau(t)) = -Q$, $Q > 0$ has a unique positive definite solution $P(\tau(t))$, for every value of $\tau(t)$. As stated, this assertion is valid only for slowly time-varying delays $\tau(t)$. Indeed, the "off the shelf" Lemma 9.9 in[19] proves that for a slow enough parameter α, if $A(\alpha)$ is differentiable and Hurwitz (uniformly in α), the Lyapunov equation has a positive definite unique solution $P(\alpha)$ for every value of α, as desired. We start by analyzing the transformed form (3.16) of the error system, using the Lyapunov–Krasovskii functional candidate [22, eq. (6.57)]:

$$V(t) = V_1(\tilde{z}) + aV_2(t), \tag{3.19}$$

$$V_1(\tilde{z}) = \tilde{z}^T P\tilde{z}, \tag{3.20}$$

$$V_2(t) = \frac{1}{2}\int_0^1 e^{b\ell}\tilde{w}^2(\ell,t)d\ell, \tag{3.21}$$

where $a > 0$ and $b > 0$ are constants to be defined later on.

Differentiating $V_1(\tilde{z})$ along the trajectories of (3.16), we can write

$$\dot{V}_1(\tilde{z}) = -\tilde{z}^T Q\tilde{z} + \tilde{z}^T P'(\tau(t))\dot{\tau}(t)\tilde{z} - 2\tilde{z}^T P(\tau(t))e^{A_a\tau(t)}L\tilde{w}(0,t), \tag{3.22}$$

where $P'(\tau(t)) = dP(\tau(t))/d(\tau(t))$. Let $\bar{p}$ be an upper bound for $\|P'(\tau(t))\|$ and define $P_{\min}$ and $P_{\max}$ as the lower and upper bounds of $\|P(\tau(t))\|$, respectively, as well as set $q \triangleq \lambda_{\min}(Q) - \bar{p}h > 0$. The positiveness expressed in the latter inequality is consistent with the condition of having a slow time-varying delay, with a sufficiently small constant h, according to assumption **(A2)**. The following inequality can be obtained:

$$\dot{V}_1(\tilde{z}) \leq -\lambda_{\min}(Q)\|\tilde{z}\|^2 + \bar{p}h\|\tilde{z}\|^2 + 2\|\tilde{z}\| P_{\max}\bar{M}\|L\||\tilde{w}(0,t)|. \tag{3.23}$$

Then, by completion of the squares, it can be shown that

$$\dot{V}_1(\tilde{z}) \leq -q\|\tilde{z}\|^2 + \frac{2(P_{\max}\bar{M}\|L\|)^2}{q}\tilde{w}^2(0,t). \tag{3.24}$$

Using (3.14) and (3.15), and integration by parts, the derivative of $V_2(t)$ can be written as

$$\begin{aligned}\dot{V}_2(t) &= \int_0^1 e^{b\ell}\tilde{w}(\ell,t)\pi(\ell,t)\tilde{w}_\ell(\ell,t)d\ell = \\ &= -\frac{\pi(0,t)\tilde{w}^2(0,t)}{2} - \frac{1}{2}\int_0^1 (b\pi(\ell,t) + \pi_\ell(\ell,t))\, e^{b\ell}\tilde{w}^2(\ell,t)d\ell.\end{aligned} \tag{3.25}$$

For the last term of (3.25) to be negative, we need

$$b\pi(\ell,t) + \pi_\ell(\ell,t) = \frac{(b(\ell-1)+1)\dot{\tau}(t) + b}{\tau(t)} > 0. \tag{3.26}$$

By assumptions **(A1)** and **(A2)**, we have $\tau(t) > 0$, $h \geq \dot{\tau}(t) \geq -h$, and $0 \leq h < 1$. Since the numerator of (3.26) is linear in ℓ, Eq. (3.25) will have its minimum at either $\ell = 0$ or $\ell = 1$, so

$$b\pi(\ell,t) + \pi_\ell(\ell,t) \geq \frac{\min\{b-h, b-h-bh\}}{\tau(t)}. \tag{3.27}$$

Choosing $b > \frac{h}{1-h} > h$, we have $b\pi(\ell,t) + \pi_\ell(\ell,t) > 0$. Then, using (3.27) in (3.25), we obtain

$$\dot{V}_2(t) \leq -\frac{\pi_0}{2}\tilde{w}^2(0,t) - \frac{\min\{b-h, b-h-bh\}}{2\tau(t)}\int_0^1 e^{b\ell}\tilde{w}^2(\ell,t)d\ell. \tag{3.28}$$

Finally, defining

$$\beta = \frac{\min\{b-h, b-h-bh\}}{1+h} > 0, \tag{3.29}$$

it is possible to verify that

$$\dot{V}_2(t) \leq -\frac{\pi_0}{2}\tilde{w}^2(0,t) - \pi_0\beta V_2(t). \tag{3.30}$$

Choosing $a = 4(P_{\max}\bar{M}\|L\|)^2/(\pi_0 q)$ and applying (3.24) and (3.30) into (3.19), one obtains the following inequality:

$$\dot{V}(t) \leq -q\|\tilde{z}\|^2 - \pi_0\beta\frac{4(P_{\max}\bar{M}\|L\|)^2}{\pi_0 q}V_2(t). \tag{3.31}$$

Defining $\mu = \min\left\{\pi_0\beta, \frac{q}{P_{\max}}\right\}$, it can be verified from the definition of $V(t)$ in (3.19) that $\dot{V}(t) \leq -\mu V(t)$, and, by a comparison lemma, that

$$V(t) \leq e^{-\mu t}V(0), \quad \forall t \geq 0. \tag{3.32}$$

This result shows that the error system $(\tilde{z}, \tilde{w})$ is exponentially stable at the origin relative to the norm

$$\Omega(t) = \|\tilde{z}(t)\|^2 + \int_0^1 \tilde{w}^2(\ell, t)d\ell. \tag{3.33}$$

Now, reminding that $P_{\min}$ and $P_{\max}$ are the lower and upper bounds of $\|P(\tau(t))\|$, respectively, it can be shown that

$$\psi_1\Omega(t) \leq V(t) \leq \psi_2\Omega(t), \tag{3.34}$$

where $\psi_1 = \min\left\{P_{\min}, \frac{2(P_{\max}\bar{M}\|L\|)^2}{q}\right\}$ and $\psi_2 = \max\left\{P_{\max}, \frac{2(P_{\max}\bar{M}\|L\|)^2}{q}e^b\right\}$. Consequently,

$$\Omega(t) \leq \frac{\psi_2}{\psi_1}e^{-\mu t}\Omega(0), \ \forall t \geq 0. \tag{3.35}$$

Based on this result, it is possible to demonstrate the stability of the system $(\tilde{z}, \tilde{v})$ in the sense of the norm

$$\Xi(t) = \|\tilde{z}(t)\|^2 + \int_0^1 \tilde{v}^2(\ell, t)d\ell, \tag{3.36}$$

by determining the relation between the norms $\Omega(t)$ and $\Xi(t)$. Utilizing the inverse *PDE-backstepping*:

$$\tilde{v}(\ell, t) = \tilde{w}(\ell, t) + Ce^{A(\ell-1)\tau(t)}\tilde{z}(t), \tag{3.37}$$

positive constants ϕ_1 and ϕ_2 can be easily determined, such that the relation between $\Omega(t)$ and $\Xi(t)$ is given by $\phi_1\Xi(t) \leq \Omega(t) \leq \phi_2\Xi(t)$. In what follows, we explicitly derive the expressions for ϕ_1 and ϕ_2 that appear in the exponential estimate (3.53).

First of all, for the sake of simplicity, let us redefine the following terms:

$$A = A_a\,, \quad B = -e^{A_a \tau(t)} L\,, \tag{3.38}$$
$$A_o = A_a - e^{A_a \tau(t)} L C_a e^{-A_a \tau(t)}\,, \quad K = C_a e^{-A_a \tau(t)}\,, \quad \phi(t) = t - \tau(t)\,.$$

Indeed, it can be shown that

$$\|\tilde{w}(t)\|_{L_2}^2 \le \alpha_1(t)\|\tilde{v}(t)\|_{L_2}^2 + \alpha_2(t)\|\tilde{z}(t)\|^2, \tag{3.39}$$
$$\|\tilde{v}(t)\|_{L_2}^2 \le \beta_1(t)\|\tilde{w}(t)\|_{L_2}^2 + \beta_2(t)\|\tilde{z}(t)\|^2, \tag{3.40}$$

where in particular we have used the compact forms $\|\tilde{w}(t)\|_{L_2} := \left(\int_0^1 \tilde{w}^2(\ell, t) d\ell\right)^{1/2}$, $\|\tilde{v}(t)\|_{L_2} := \left(\int_0^1 \tilde{v}^2(\ell, t) d\ell\right)^{1/2}$ for the L_2-norm of the PDE state variables, and

$$\alpha_1(t) = 3\left(1 + \int_0^1 (KM(x(\phi^{-1}(t) - t))B(\phi^{-1}(t) - t))^2 dx\right), \tag{3.41}$$

$$\alpha_2(t) = 3\int_0^1 \|KM(x(\phi^{-1}(t) - t))\|^2 dx, \tag{3.42}$$

$$\beta_1(t) = 3\left(1 + \int_0^1 (KN(x(\phi^{-1}(t) - t))B(\phi^{-1}(t) - t))^2 dx\right), \tag{3.43}$$

$$\beta_2(t) = 3\int_0^1 \|KN(x(\phi^{-1}(t) - t))\|^2 dx, \tag{3.44}$$

and

$$M(s) = e^{As}\,, \quad N(s) = e^{A_o s}\,, \quad \pi_0^* = \frac{1}{\sup_{\vartheta \ge \phi^{-1}(0)} (\vartheta - \phi(\vartheta))} > 0\,. \tag{3.45}$$

Furthermore, we can show that

$$\alpha_1(t) \le \bar{\alpha}_1 = 3\left(1 + \|K\|^2 \|B\|^2 \frac{e^{\frac{2\|A\|}{\pi_0^*}} - 1}{2\pi_0^* \|A\|}\right), \tag{3.46}$$

$$\alpha_2(t) \le \bar{\alpha}_2 = 3\|K\|^2 \pi_0^* \frac{e^{\frac{2\|A\|}{\pi_0^*}} - 1}{2\|A\|}, \tag{3.47}$$

$$\beta_1(t) \le \bar{\beta}_1 = 3\left(1 + \|K\|^2 \|B\|^2 \frac{e^{\frac{2\|A_o\|}{\pi_0^*}} - 1}{2\pi_0^* \|A_o\|}\right) \tag{3.48}$$

$$\beta_2(t) \leq \bar{\beta}_2 = 3\|K\|^2 \pi_0^* \frac{e^{\frac{2\|A_o\|}{\pi_0^*}} - 1}{2\|A_o\|}. \tag{3.49}$$

With a few substitutions, we obtain

$$\phi_1 \Xi(t) \leq \Omega(t) \leq \phi_2 \Xi(t), \tag{3.50}$$

where

$$\phi_1 = \frac{1}{\max\{\bar{\beta}_1 , \bar{\beta}_2 + 1\}}, \tag{3.51}$$

$$\phi_2 = \max\{\bar{\alpha}_1 , \bar{\alpha}_2 + 1\}. \tag{3.52}$$

Consequently,

$$\Xi(t) \leq \frac{\phi_2 \psi_2}{\phi_1 \psi_1} e^{-\mu t} \Xi(0), \tag{3.53}$$

showing that the error system $(\tilde{z}, \tilde{v})$ is stable at the origin in the sense of the norm (3.17). □

It is worth mentioning that the PDE-backstepping transformation (3.13) and its inverse (3.37) are bounded since they are constructed with the error variables (3.11) that are proved to converge to zero exponentially in Theorem 3.1, and, analogously, in prescribed time in the next Theorem 3.3, as shown in [11, Theorem 1].

3.2.3 Alternative Representations of the Observer

Based on (3.10), it can be shown, as in [22, Cor. 3.1], that the observer can be represented in terms of the output as

$$\begin{cases} \dot{\hat{z}}(t) = A_a \hat{z}(t) + B_a u(t) + e^{A_a \tau(t)} L \left(y(t) - \hat{y}(t)\right) \\ \hat{y}(t) = C_a \left(\hat{z}(t - \tau(t)) + \int_{t-\tau(t)}^{t} e^{A_a(t-\theta)} L \left(y(\theta) - \hat{y}(\theta)\right) d\theta\right) \end{cases} . \tag{3.54}$$

Alternatively, as in [22, Section 6.3] and in [24], if the input signal $u(t)$ has its past values known for $t \in [t - \tau, t]$, it can be shown that the representation

$$\begin{cases} \dot{\zeta}(t) = (1 - \dot{\tau}(t)) \left(A_a \zeta(t) + B_a u(t - \tau(t)) + L \left(y(t) - C_a \zeta(t)\right)\right) \\ \hat{z}(t) = e^{A_a \tau(t)} \zeta(t) + \int_{t-\tau(t)}^{t} e^{A_a(t-\theta)} B_a u(\theta) d\theta, \end{cases} \tag{3.55}$$

where $\zeta = \hat{z}(t - \tau(t))$, is equivalent to both the PDE-based (3.10) and the output-based (3.54) versions. Notice that this representation consists of an observer for the delayed system cascaded with a feedforward predictor based on the *variation of constants* formula[1] [20, 43].

Such alternative versions are easier to implement numerically. On the other hand, the ODE-PDE version helps to prove the stability by enabling a quantification of exponential stability in a suitable norm for the closed-loop system, which is infinite dimensional—see the proof of Theorem 3.1. For more details, see also Chapters 2 and 3 of reference [22].

3.3 Sliding-Mode Control Design

For the results of this section, we need to invoke Filippov's definition for solution of discontinuous differential equations [12] due to the existence of the sliding modes and differential equations with discontinuous right-hand sides.

In the augmented system (3.5), only the output signal is available, presenting a measurement delay. In order to design a sliding-mode control, we can use estimates for the current state of the system, obtained by the observer (3.10), to define the sliding variable:

$$\hat{\sigma} = S\hat{z}, \tag{3.56}$$

where $\hat{\sigma} \in \mathbb{R}$, and $S \in \mathbb{R}^{1\times(n+n_d)}$ is a full-rank matrix that will be chosen to design the sliding surface $\mathcal{S} = \{\hat{z} : S\hat{z} = 0\}$. Also, the following auxiliary variables are defined:

$$\sigma = Sz, \quad \tilde{\sigma} = S\tilde{z}. \tag{3.57}$$

From (3.11), it can be seen that

$$\tilde{\sigma} = \sigma - \hat{\sigma}. \tag{3.58}$$

[1] The *variation of constants* formula is also known as the method of variation of parameters, the Lagrange's method, or Duhamel's principle. More specifically in PDEs, Duhamel's principle is a general method for obtaining solutions to inhomogeneous linear evolution equations. For linear evolution equations without spatial dependency, Duhamel's principle reduces to the method of variation of parameters technique for solving linear inhomogeneous ordinary differential equations (ODEs). The philosophy underlying Duhamel's principle is that it is possible to go from solutions of the Cauchy problem (or initial value problem) to solutions of the inhomogeneous problem.

To choose the matrix S, consider the augmented equations (3.5) and apply the representation for the system in (3.3):

$$\begin{cases} \dot{x}_d(t) = A_d x_d(t) \\ \dot{x}_1(t) = A_{11}x_1(t) + A_{12}x_2(t) + F_1 C_d x_d(t) \\ \dot{x}_2(t) = A_{21}x_1(t) + A_{22}x_2(t) + F_2 C_d x_d(t) + B_2 u(t) \end{cases} . \tag{3.59}$$

3.3.1 A Case of Relative Degree One

Assume $C = [\mathbf{0}^{1\times(n-1)} \quad C_2]$. Then, note that the relative degree from u to x_2 is uniformly unitary since B_2 is nonzero by assumption **(A3)**. An SMC law can be designed to regulate x_2 to zero consequently regulating the output y to zero. The natural choice for S is thus

$$S = \left[\mathbf{0}^{1\times(n+n_d-1)} \ 1\right], \tag{3.60}$$

and then, $\hat{\sigma} = S\hat{z} = \hat{x}_2(t)$. When a sliding mode is reached in $\mathcal{S}$, the sliding variable is bound to $\hat{\sigma} = 0$. Then, considering (3.57) and (3.58), we have $\hat{\sigma} = S\hat{z} = Sz - \tilde{\sigma} = 0$, which, by substitution of S, leads to

$$\hat{x}_2(t) = x_2(t) - \tilde{\sigma} = 0. \tag{3.61}$$

From Theorem 3.1, it can be seen that $\tilde{\sigma} = S\tilde{z}$ is an exponentially decaying term. Therefore, x_2 will converge exponentially to zero, and, in consequence, $y(t) = C_2 x_2(t - \tau(t))$ will also be brought to zero, hence rejecting the disturbance.

To determine the control law, it follows from (3.54) that

$$\begin{aligned} \dot{\hat{\sigma}} &= S\dot{\hat{z}}(t) = SA_a\hat{z}(t) + SB_a u(t) + SL_\tau(t)\left(y(t) - \hat{y}(t)\right), \\ L_\tau(t) &\triangleq e^{A_a\tau(t)}L, \quad \|L_\tau(t)\| \leq \bar{M}\|L\|. \end{aligned} \tag{3.62}$$

From assumption **(A3)**, it can be verified that, since the chosen matrix S is full rank, matrix SB_a is nonsingular. Then, applying the equivalent control method [8, 32], it is possible to obtain the equivalent control u_{eq} from the solution of the equation $\dot{\hat{\sigma}} = 0$:

$$\begin{aligned} &SA_a\hat{z}(t) + SB_a u_{\text{eq}}(t) + SL_\tau(t)\left(y(t) - \hat{y}(t)\right) = 0, \\ &u_{\text{eq}}(t) = -(SB_a)^{-1}SA_a\hat{z}(t) - (SB_a)^{-1}SL_\tau(t)\left(y(t) - \hat{y}(t)\right). \end{aligned} \tag{3.63}$$

Adding a stabilizing linear term and a discontinuous term that drives the state to the sliding surface, the proposed control law is then given by

$$u(t) = u_{\text{eq}}(t) - (SB_a)^{-1}k\hat{\sigma} - (SB_a)^{-1}\rho(t)\,\text{sgn}(\hat{\sigma}), \tag{3.64}$$

where $k > 0$ is the gain of the stabilizing term and the modulation $\rho(t)$ is a scalar positive function, bounded by

$$\rho(t) \geq \eta, \tag{3.65}$$

where $\eta > 0$ is any arbitrary constant. Also note that if $\rho(t) = 0$ is used, a linear control law that will also reject the disturbance is obtained. Indeed, the second term in the right-hand side of (3.64) contains $\hat{z}(t)$ since $\hat{\sigma}(t) = S\hat{z}(t)$, and consequently, the portion dedicated to cancel the external disturbance.

The discontinuous term in (3.64) gives the finite-time stabilization of the sliding variable. In addition, this switching term brings robustness with respect to parameter uncertainties for the closed-loop system, as illustrated in the simulation results of Sect. 3.5.1. At last, with the switching term, we open the door to consider switched actuators, which are very well motivated in power electronics and aerospace applications.

Moreover, the equivalent control term can be useful in order to reduce the real-world chattering from the discontinuous control term since a smaller modulation function (or control gain) $\rho(t)$ can be considered. The other option would be to incorporate the equivalent control term into the function $\rho(t)$ of (3.64), leading to a single discontinuous control law with increased amplitudes, and, consequently, more control chattering [42].

3.3.2 Closed-Loop Stability Analysis

Theorem 3.2 *Consider the delayed output system (3.1), under assumptions **(A1)** to **(A6)**, in the presence of a disturbance $d(t)$ modeled by a linear dynamic exosystem (3.4), under assumptions **(A7)** and **(A8)**. Using the state estimate for system (3.5) obtained from the observer-predictor (3.54) to construct the sliding variable $\hat{\sigma} = \hat{x}_2$ and the control law (3.64), the following properties hold:*

1. *An ideal sliding mode $\hat{\sigma}(t) \equiv 0$ is reached in finite time, no greater than $|\hat{\sigma}(0)|/\eta$.*
2. *The state x_2, and thus the output signal $y(t) = C_2 x_2$ is exponentially convergent to zero, for all initial conditions.*
3. *All signals of the closed-loop system remain uniformly bounded.*

Proof Applying the control law (3.64) in (3.62), one obtains

$$\dot{\hat{\sigma}} = -k\hat{\sigma} - \rho(t)\,\mathrm{sgn}(\hat{\sigma}). \tag{3.66}$$

Using the Lyapunov function candidate $V_3(\hat{\sigma}) = (\hat{\sigma}^T\hat{\sigma})/2$, it can be verified that

$$\dot{V}_3(\hat{\sigma}) = \hat{\sigma}^T\dot{\hat{\sigma}} \leq -k|\hat{\sigma}|^2 - \rho(t)|\hat{\sigma}|. \tag{3.67}$$

Then, since $k > 0$ and substituting (3.65), we obtain

$$\dot{V}_3(\hat{\sigma}) \leq -\eta|\hat{\sigma}|. \tag{3.68}$$

From the definition of V_3, we can observe that

$$\dot{V}_3(\hat{\sigma}) \leq -\sqrt{2}\eta\sqrt{V_3(\hat{\sigma})}, \tag{3.69}$$

which shows that the sliding surface is reached in some finite time $t_f < |\hat{\sigma}(0)|/\eta$ and remains there for all future time, i.e., $\hat{\sigma}(t) \equiv 0$, $\forall t \geq t_f$, which shows that the sliding mode is ideal.

From Theorem 3.1, it can be seen that $\tilde{\sigma} = S\tilde{z}$ is an exponentially decaying term. Considering (3.57) and (3.58), it can be verified that $\sigma \to 0$ exponentially. Therefore, x_2 will converge exponentially to zero, and, since $y(t) = C_2 x_2(t - \tau(t))$, the output will also be taken to zero, rejecting the disturbance.

Then, the system (3.59) will be reduced to the following form:

$$\begin{cases} \dot{x}_d(t) = A_d x_d(t) \\ \dot{x}_1(t) = A_{11} x_1(t) + F_1 C_d x_d(t) . \end{cases} \tag{3.70}$$

Since, by assumption **(A6)**, the system is minimum phase, A_{11} is Hurwitz, and that, by assumption **(A8)**, $x_d(t)$ is bounded, then it can be seen that the state x_1 is bounded. Also, from Theorem 3.1, it can be concluded that since x_1 and x_d are bounded, their estimates $\hat{x}_1$ and $\hat{x}_d$ are also bounded. □

The results above are relevant even for constant delays—the constant-delay case had not been previously addressed in the literature yet. The strength of Theorem 3.2 is that, while predictors are not available explicitly under nonlinear control in general because usually it is not possible to solve nonlinear systems explicitly, when the plant is linear, we have an explicit predictor, even though the SMC is nonlinear.

3.4 Prescribed-Time Estimation and Finite-Time Disturbance Rejection for Constant Delays

For the results of this section, we need to invoke *generalized* Filippov solutions [31] due to the existence of the sliding modes as well as the mathematical technicalities involving a right-hand side exhibiting singularity at the prescribed convergence time instant.

In this section, we consider systems represented by (3.5), but subject to *constant* time delays $D > 0$ in the output signal, i.e.,

$$\begin{cases} \dot{z}(t) = A_a z(t) + B_a u(t) \\ y(t) = C_a z(t - D), \end{cases} \tag{3.71}$$

with $t \geq t_0 \geq 0$, and the initial condition being defined by $z(\theta) = z_0(\theta)$ for all $\theta \in [t_0 - D, t_0]$, where t_0 is the initialization time, $u(t) \in \mathbb{R}$ is the input signal, and $y(t) \in \mathbb{R}$ is the delayed output signal.

We recall that the observability matrix is given by

$$\mathscr{O} = \left[C_a^T \ A_a^T C_a^T \cdots \left(A_a^{n+n_d-1}\right)^T C_a^T \right]^T ,$$

where we assume that (A_a, C_a) is an observable pair, i.e., that $\text{rank}(\mathscr{O}) = n + n_d$. We perform the coordinate transformation [11]

$$z_{ob}(t) = Q^{-1} z(t) , \tag{3.72}$$

where $Q = \left[\tilde{q}, A_a\tilde{q}, \ldots, A_a^{n+n_d-1}\tilde{q}\right]$, with $\tilde{q}$ being the $(n+n_d)$-th column of $\mathscr{O}^{-1}$. An application of (3.72) to (3.71) transforms it into the following system in the observer form [2]:

$$\begin{cases} \dot{z}_{ob}(t) = A_{ob} z_{ob}(t) + B_{ob} u(t), \\ y_{ob}(t) = C_{ob} z_{ob}(t - D), \end{cases} \tag{3.73}$$

where $A_{ob} = Q^{-1} A_a Q$, $C_{ob} = C_a Q$, and $B_{ob} = Q^{-1} B_a$. More precisely,

$$A_{ob} = \begin{bmatrix} 0 & 0 & 0 & \cdots & -a_0 \\ 1 & 0 & 0 & \cdots & -a_1 \\ \vdots & \ddots & & \cdots & \vdots \\ 0 & 0 & \cdots & 1 & -a_{n+n_d-1} \end{bmatrix}, \quad C_{ob} = \begin{bmatrix} 0 & \ldots & 1 \end{bmatrix}. \tag{3.74}$$

Our goal is to design an observer for the system (3.71), which converges within the prescribed terminal time. To this end, we employ the methodology developed in [11], which relies on representing the sensor delay as a linear transport PDE and builds upon the cascade ODE-PDE setting of the reference [22, Chapter 3]. We henceforth represent system (3.71) as

$$\begin{cases} \dot{z}(t) &= A_a z(t) + B_a u(t), \\ u_t(t, \ell) &= u_\ell(t, \ell), \\ u(t, D) &= C_a z(t), \\ y(t) &= u(t, 0), \end{cases} \tag{3.75}$$

where $t \geq t_0 \geq 0$, $\ell \in [0, D]$, and $u(t, \cdot)$ is the transport PDE state at time t whose solution is $u(t, \ell) = C_a z(t - D + \ell)$, and, in particular, $u(t, 0) = y(t)$.

Under the change of coordinates (3.72), system (3.73) rewrites as

$$\begin{cases} \dot{z}_{ob}(t) &= A_{ob} z_{ob}(t) + B_{ob} u(t), \\ u_t(t,\ell) &= u_\ell(t,\ell), \\ u(t,D) &= C_{ob} z_{ob}(t). \\ y_{ob}(t) &= u(t,0). \end{cases} \tag{3.76}$$

As shown in [11], the above observer form reformulation will be instrumental in performing several other transformations to achieve a full delay-compensated observer design, which we discuss in the next subsection.

On the other hand, now the control objective is to guarantee a finite-time disturbance rejection scheme by means of prescribed-time state estimation—the disturbance to be rejected is still that given in Eq. (3.4). This inclusion is extremely important since in Sect. 3.3 only asymptotic results could be proved.

3.4.1 *Delay-Compensated Prescribed-Time Observer*

In this section, it is presented an observer that converges in the prescribed terminal time $t_0 + D + T$, where $T \in \mathbb{R}_+$ is a priori fixed, D is the known sensor delay, and t_0 is the initialization time.

To achieve prescribed-time convergence, the authors in [11] have used time-varying observer gains constructed with suitable blow-up functions. The approach builds upon the recent ones for prescribed-time stabilization and estimation of ODEs [9, 16, 37], which proposes predictor feedback for prescribed-time input delay compensation of LTI systems.

Consider the following blow-up function (as in [10]):

$$c(t) = \frac{\bar{c}_0^2 T^2}{(T + t_0 - t)^2}, \quad c(t_0) = \bar{c}_0^2, \tag{3.77}$$

with $\bar{c}_0 > 0$. As discussed in [9] and [10], the chosen power degree of the blow-up function (3.77) is crucial for the convergence analysis; moreover, it allows us to obtain explicit and compact closed-form gain formulas throughout the design/analysis.

As in [11], we consider the following time-varying matrices $P(\cdot)$ and $M(\cdot)$ in companion canonical/observer form, respectively:

$$P(t) = \begin{bmatrix} 0 & 1 & 0 & \cdots & 0 \\ \vdots & \vdots & \vdots & \ddots & \vdots \\ 0 & 0 & 0 & \cdots & 1 \\ -p_0(t) & -p_1(t) & -p_2(t) & \cdots & -p_{n+n_d-1}(t) \end{bmatrix}, \tag{3.78}$$

$$M(t) = \begin{bmatrix} 0 & 0 & 0 & \cdots & -\mu_0(t) \\ 1 & 0 & 0 & \cdots & -\mu_1(t) \\ \vdots & \ddots & & \cdots & \vdots \\ 0 & 0 & \cdots & 1 & -\mu_{n+n_d-1}(t) \end{bmatrix}, \tag{3.79}$$

where $p_{i-1}(\cdot)$, $\mu_{i-1}(\cdot)$, and $i = 1 \ldots n + n_d$ are time-varying functions that will be characterized below. There exists a time-varying matrix $R(\cdot)$ such that

$$R(t)M(t) = P(t)R(t) - \dot{R}(t)\,. \tag{3.80}$$

Indeed, R is given as follows [34]:

$$R(t) = \begin{bmatrix} 0 & 0 & \cdots & 0 & 1 \\ 0 & 0 & \cdots & 1 & \pi_{n+n_d-1,n+n_d-1}(t) \\ \vdots & \vdots & & \vdots & \vdots \\ 1 & \pi_{1,1}(t) & \cdots & \pi_{n+n_d-2,1}(t) & \pi_{n+n_d-1,1}(t) \end{bmatrix}, \tag{3.81}$$

where the time-varying coefficients $\pi_{i,k}(\cdot)$ are characterized by the following recurrence relation:

$$\pi_{i,k} = \begin{cases} -p_{n+n_d-1}, & 1 \le k = 1 \le n + n_d, \\ \pi_{i-1,k-1} - \dot{\pi}_{i-1,k}, & 1 < k < i \le n + n_d, \\ -p_{n+n_d-i} - \sum_{j=0}^{i-2} p_{n+n_d-1-j}\pi_{i-1,j+1} - \dot{\pi}_{i-1,1}, & k = 1 < i \le n + n_d. \end{cases} \tag{3.82}$$

Hence, from (3.80) along with (3.78), (3.79), and (3.81), the time-varying coefficients of $M(t)$ can be explicitly characterized as follows [40, Section 2], [26]:

$$\mu_k(t) = \sum_{i=k}^{n+n_d-1} (-1)^{i-k} \binom{i}{k} p_i^{(i-k)}(t), \quad k = 0, \ldots, n + n_d - 1, \tag{3.83}$$

where $\binom{i}{k} := \frac{i!}{k!(i-k)!}$ denotes the binomial coefficients.

Under the same assumptions of Proposition 1 of reference [9, Section 2], the entries $p_0(t)$, $p_j(t)$ $(j = 1, \cdots, n + n_d - 1)$ in (3.78) are explicitly characterized by

$$p_0(t) = \sigma_{n+n_d}(r_1, \cdots, r_{n+n_d})c^{n+n_d}(t), \tag{3.84}$$

and for $j = 1, \cdots, n + n_d - 1$,

$$p_j(t) = \frac{\left(\sqrt{c(t)}\right)^{n+n_d-j}}{(\bar{c}_0 T)^{n+n_d-j}} \sum_{k=j}^{n+n_d} (-1)^{k-j} \sigma_{n+n_d-k}(r_1, \cdots, r_{n+n_d}) \times \binom{k-1}{j-1} \frac{k!}{j!} \left(\bar{c}_0 T \sqrt{c(t)}\right)^{n+n_d-k}, \tag{3.85}$$

where $c(t)$ is given in (3.77), $\sigma_{n+n_d-k}(\cdot)$ are the elementary symmetric polynomials given by

$$\sigma_0(r_1, \cdots, r_{n+n_d}) = 1, \tag{3.86}$$

$$\sigma_k(r_1, \cdots, r_{n+n_d}) = \sum_{1 \le i_1 \le i_2 \le \cdots i_k \le n+n_d} r_{i_1} r_{i_2} \cdots r_{i_k}, \tag{3.87}$$

$$\sigma_k(r_1, \cdots, r_{n+n_d}) = \prod_{i=1}^{n+n_d} r_i, \tag{3.88}$$

and $\sigma_k(r_1, \cdots, r_{n+n_d}) = 0$, for $k > n + n_d$. The coefficients r_i, $i = 1, \ldots, n + n_d$, are simply distinct positive real numbers (*i.e.*, $r_i > 0$, $r_i \neq r_j$ for $i \neq j$, for $i = 1, \ldots, n + n_d$).

Hence, the prescribed-time observer is given in the following theorem. It is an infinite-dimensional, full-order observer as it estimates both the plant state $z(t)$ and the sensor state $y(t)$ (*i.e.*, $y(t)$ is introduced in the estimation error).

Theorem 3.3 ([11, Theorem 1]) *Let Q be as in (3.72), a_{i-1} be as in (3.74), $c(t - D)$ be defined according to (3.77), $\mu_{i-1}(t - D)$ be given by (3.83), (3.84)–(3.85) (for $i = 1, \cdots, n$), and let $T > 0$ be fixed. For the system (3.75), the observer*

$$\begin{cases} \dot{\hat{z}}(t) = A_a \hat{z}(t) + B_a u(t) + Q e^{Q^{-1} A_a Q D} \gamma(t - D) \left(y(t) - \hat{u}(t, 0)\right) \\ \hat{u}_t(t, \ell) = \hat{u}_\ell(t, \ell) + C_a Q e^{Q^{-1} A_a Q \ell} \gamma(t - D) \left(y(t) - \hat{u}(t, 0)\right) \\ \hat{u}(t, D) = C_a \hat{z}(t) \end{cases}, \tag{3.89}$$

with time-varying output gain $\gamma(t - D) = \left[\gamma_1(t - D), \ldots, \gamma_{n+n_d}(t - D)\right]^T$ given by

$$\gamma_i(t - D) = -a_{i-1} + \mu_{i-1}(t - D), \quad for \quad i = 1, \cdots, n + n_d, \tag{3.90}$$

guarantees that $(\hat{z}, \hat{u})$ converges to (z, u) within the prescribed terminal time $t_0 + T + D$. More precisely, there exist a positive constant $\bar{\zeta}_z$ and a positive polynomial $\mathscr{P}(\sqrt{c(t - D)})$ such that, for any initial conditions $z(t_0)$, $\hat{z}(t_0) \in \mathbb{R}^{n+n_d}$, and for any initial measurements $u(t_0, \ell) = C_a z(t_0 - D + \ell)$, $\hat{u}(t_0, \ell) = C_a \hat{z}(t_0 - D + \ell)$, *where* $\ell \in [0, D]$, *we have that the quantity* $\|z(t) - \hat{z}(t)\|^2 + \|u(t,\cdot) - \hat{u}(t,\cdot)\|_{L_2}^2$ *remains bounded for* $t \in [t_0, t_0 + D]$, *whereas it satisfies the bound*

$$\|z(t) - \hat{z}(t)\|^2 + \|u(t,\cdot) - \hat{u}(t,\cdot)\|_{L_2}^2 \leq \bar{\xi}_z \mathscr{P}\left(\sqrt{c(t-D)}\right)$$
$$\exp\left(-r_{\min}\bar{c}_0 T \sqrt{c(t-D)}\right) \times \|z(t_0) - \hat{z}(t_0)\|^2 \tag{3.91}$$

for $t \in [t_0 + D, t_0 + T + D]$ *with* $r_{\min} = \min_{i=1,\ldots,n+n_d}\{r_i\}$, *and in particular,*

$$\|z(t) - \hat{z}(t)\|^2 + \|u(t,\cdot) - \hat{u}(t,\cdot)\|_{L_2}^2 \to 0 \tag{3.92}$$

as $t \to t_0 + T + D$, *where* $\|u(t,\cdot) - \hat{u}(t,\cdot)\|_{L_2}^2 := \int_0^D \left(u(t,\ell) - \hat{u}(t,\ell)\right)^2 d\ell$.

The Lyapunov analysis carried out in [11, Theorem 1] to prove Theorem 3.3 does quantitatively capture the dependence on time and on the initial conditions in the chosen norm (note that this choice is not unique). Hence, stability is uniform in the initial time. This is explicitly captured in the KL estimates in (76). It is achieved by incorporating the initial time in the time-varying gains.

The presented observer has several distinctive features: The observer gains (3.90) are explicit; the terminal time $t = t_0 + T + D$ can be a priori prescribed regardless of the size of initial conditions; the convergence of the estimate errors is highly smooth [11, Remark 3]; and the observer estimates both the plant states and the sensor state.

One particular aspect of this section is to note that the augmented system (3.5) and Eq. (3.71) have the same form when $\tau(t) = D$ and then employ the prescribed-time observer [11] governed by (3.89) in closed-loop feedback through the SMC law introduced before in Sect. 3.3, as described in the following.

3.4.2 Closed-Loop Stability Analysis

Theorem 3.4 *Consider the delayed output system (3.1) with a constant time delay* $\tau(t) := D > 0$, *under assumptions* ***(A1)*** *to* ***(A6)***, *in the presence of a disturbance* $d(t)$ *modeled by a linear dynamic exosystem (3.4), under assumptions* ***(A7)*** *and* ***(A8)***. *Using the state estimate for system (3.5), with* $\tau(t) = D$, *obtained from the observer-predictor of the form (3.89), with time-varying gain* $\gamma(t - D)$ *given in (3.90) to construct the sliding variable* $\hat{\sigma} = \hat{x}_2$ *and the control law (3.64), the following properties hold:*

1. *An ideal sliding mode* $\hat{\sigma}(t) \equiv 0$ *is reached in finite time, no greater than* $|\hat{\sigma}(t_0)|/\eta$.
2. *The estimation error* $z(t) - \hat{z}(t) \to 0$ *within the prescribed terminal time* $t_0 + T + D$, *for* $T > |\hat{\sigma}(t_0)|/\eta$.

3. *The state x_2, and thus the output signal $y(t) = C_2 x_2$ is convergent to zero at least in finite time, for all initial conditions.*
4. *All signals of the closed-loop system remain uniformly bounded.*

Proof Replacing the exponential observer (3.10) by the prescribed-time version given in (3.89) to estimate $\hat{z}(t)$, we continue to be able to apply the control law (3.64) in (3.62) and obtain

$$\dot{\hat{\sigma}} = -k\hat{\sigma} - \rho(t)\,\text{sgn}(\hat{\sigma}). \tag{3.93}$$

In this sense, the same Lyapunov function candidate $V_3(\hat{\sigma}) = (\hat{\sigma}^T\hat{\sigma})/2$ considered in the proof of Theorem 3.2 can be employed to conclude that

$$\dot{V}_3(\hat{\sigma}) \leq -\sqrt{2}\eta\sqrt{V_3(\hat{\sigma})}. \tag{3.94}$$

Thus, the sliding surface is reached in some finite time $t_f < |\hat{\sigma}(t_0)|/\eta < T$ and remains there for all future time, *i.e.*, $\hat{\sigma}(t) \equiv 0$, for $t_f \leq t < T$. From Theorem 3.3, it can be seen that $\tilde{\sigma} = S\tilde{z}$ is a prescribed-time decaying term. Considering (3.57) and (3.58), it can be verified that $\sigma \to 0$ in finite time since $\sigma(t) = \tilde{\sigma}(t) + \hat{\sigma}(t)$. Therefore, $x_2(= \sigma)$ will converge to zero in finite time, and, since $y(t) = C_2 x_2(t - \tau(t))$, the output will also be driven to zero in finite time, by rejecting the disturbance in finite time as well. Again, the closed-loop system (3.59) will be reduced to (3.70), which is minimum phase. Since $x_d(t)$ is bounded by construction, it can be concluded that the state x_1 is also bounded. Additionally, from Theorem 3.3, it can be concluded that the estimates $\hat{x}_1$ and $\hat{x}_d$ are bounded since x_1 and x_d are bounded as well. □

Note that prescribed-time disturbance rejection would be possible if the control gain $\rho(t)$ in the proposed SMC law (3.64) was also designed in terms of the blow-up function (3.77). Although such a result seems to be interesting for future investigation, it may result in larger control efforts. In this sense, the focus on this section was to guarantee finite-time convergence results (faster than exponential) without sacrificing the control signal. Indeed, most existing finite-time control methods are based on fractional or SMC in that the control schemes are of the form $u = -c\text{sgn}(x)|x|^\alpha$ with $c > 0$ and $0 < \alpha < 1$, essentially signum function based when the fraction $\alpha = 0$, *i.e.*, $u = -c\text{sgn}(x)$. Another possibility is the time-varying feedback alternative to achieving finite-time control, which is referred to as prescribed-time control [38] or predefined-time control [18].

3.5 Simulations

3.5.1 *Asymptotic Disturbance Rejection with Exponential Predictor*

In order to demonstrate the effectiveness of the proposed strategy, a mass-spring-damper system, made intentionally unstable by selecting a negative damping factor, with an actuation force as the control input will be used. The disturbance is a sinusoidal signal $d(t) = -(\pi/8)\sin(10t)$ fed to the velocity equation ($\dot{x}_1$), so that the disturbance signal is completely unmatched from the input signal. The system presents a time-varying measurement delay $\tau(t) = 1 + 0.1\sin(t)$.

The matrices of the state representation for the system and the disturbance are

$$A = \begin{bmatrix} 0 & 1 \\ -2 & 1 \end{bmatrix},\ B = \begin{bmatrix} 0 \\ 2 \end{bmatrix},\ F = \begin{bmatrix} 1 \\ 0 \end{bmatrix},\ A_d = \begin{bmatrix} 0 & 1 \\ -100 & 0 \end{bmatrix}, C_d = \begin{bmatrix} 0 & 0.05 \end{bmatrix},$$
$$C = \begin{bmatrix} 0 & 1 \end{bmatrix}. \tag{3.95}$$

The observer-predictor was implemented as in (3.55), which has less computation complexity, with L chosen such that the eigenvalues of $(A_a - LC_a)$ were $\{-12, -11, -10, -9\}$. A sliding-mode control law as proposed in (3.64) was designed with $S = \begin{bmatrix} 0 & 0 & 0 & 1 \end{bmatrix}$, a linear stabilizing gain $k = 1$, and a constant modulation (gain) $\rho(t) = 0.5$.

In Fig. 3.1, we compare three control laws: 1) The SMC law from (3.64); 2) a linear law obtained from (3.64) by setting $\rho(t) = 0$; 3) a linear feedback law $u = -K\hat{x} - K_d\hat{x}_d$ as described in [3, eq (4.22)]. The corresponding control signals are shown in Fig. 3.2. The latter results in

$$\dot{x} = (A - BK)x + (FC_d - BK_d)x_d + \pi_z, \tag{3.96}$$

where π_z is an exponentially decaying term due to the observation error.

According to [3], the disturbance distribution matrix $(FC_d - BK_d)$ should be minimized. However, in this particular case, such norm is minimized by $K_d =$

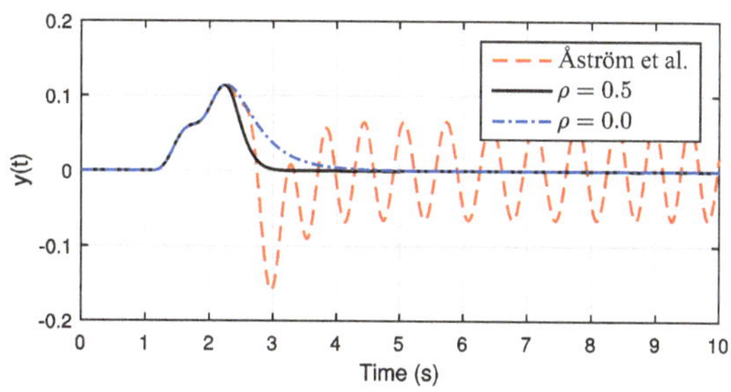

Fig. 3.1 Comparison of the output signal for the linear feedback controller from [3] (dashed/red), the control law proposed in (3.64) with $\rho = 0.5$ (solid/black), and the linear control law obtained from (3.64) using $\rho = 0$ (dash-dot/blue)

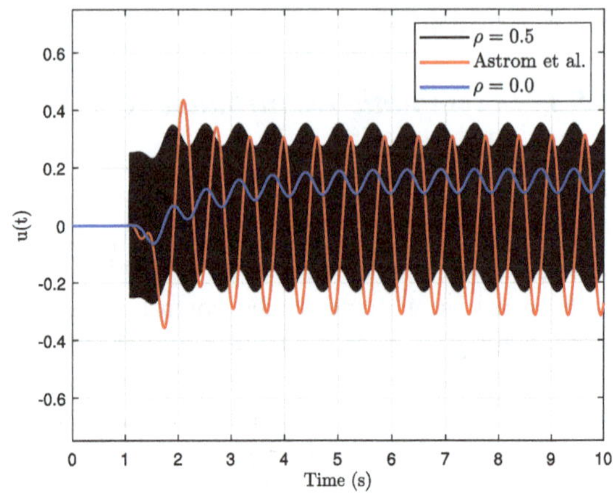

Fig. 3.2 Comparison of the control signal for the linear feedback controller from [3] (red), the control law proposed in (3.64) with $\rho = 0.5$ (black), and the linear control law obtained from (3.64) using $\rho = 0$ (blue)

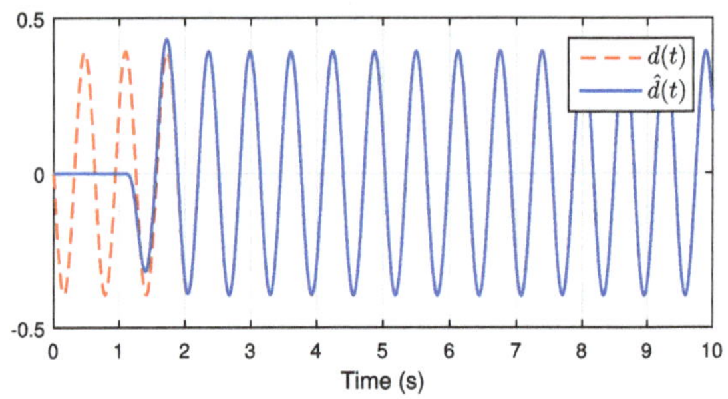

Fig. 3.3 Sinusoidal disturbance signal (dashed) and its estimate (solid)

$\begin{bmatrix} 0 & 0 \end{bmatrix}$, since F and B are orthogonal. Both the sliding-mode and linear laws based on (3.64) were able to completely reject the disturbance, with the sliding-mode law being faster than the linear law. In contrast with the control law [3], disturbance rejection was not possible.

Figure 3.3 compares the disturbance signal and its estimate obtained by the observer-predictor. After the initial delay time, the predicted output matches the system states, and the disturbance can be reconstructed.

In order to verify the performance of the closed-loop system, the uncertainties

$$\delta A = \begin{bmatrix} 0 & 0 \\ .5 & .5 \end{bmatrix}, \; \delta B = \begin{bmatrix} 0 \\ .2 \end{bmatrix}, \; \delta A_d = \begin{bmatrix} 0 & 0 \\ 5 & 0 \end{bmatrix}, \tag{3.97}$$

were added to the system matrices A, B, and A_d, respectively, while the observer and control law were designed considering their nominal value (3.95).

Figure 3.4 shows the comparison of the closed-loop output signal for the system presenting the uncertainties above. The sliding-mode control law proposed in (3.64) with $k = 2$ and $\rho(t) = 0.5$ (solid line/black) was able to bring the output to a small residual error, while the linear control law obtained from (3.64) using the same linear gain, but $\rho(t) = 0$ (dash-dot line/blue) was unable to stabilize the

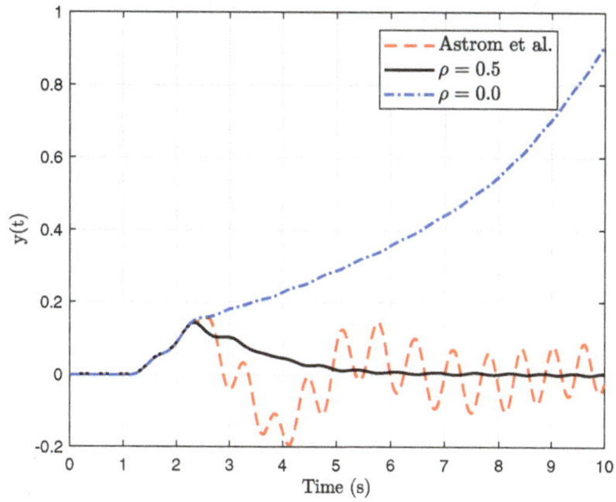

Fig. 3.4 Comparison of the closed-loop output for a system with uncertainties in matrices A, B, and A_d, using: the proposed sliding-mode control law (3.64) with $\rho = 0.5$ (solid/black), its linear counterpart with $\rho = 0$ (dash-dot/blue), and the linear feedback controller from [3] (dashed/red). The linear controllers failed to stabilize the output in the presence of uncertainties

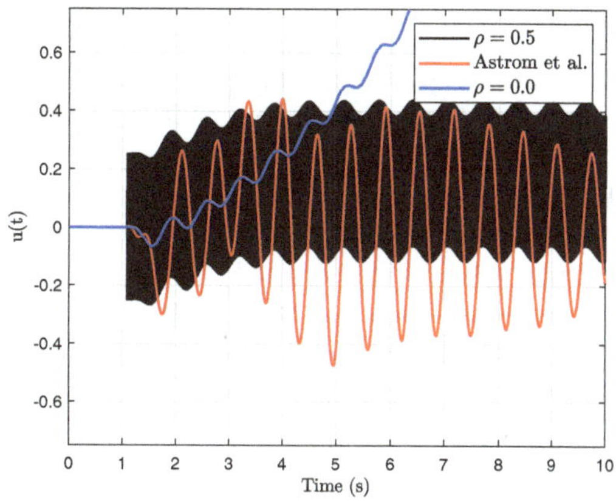

Fig. 3.5 Comparison of the control signals for the closed-loop system presenting uncertainties in matrices A, B, and A_d, using the sliding-mode control law proposed in (3.64) with $\rho = 0.5$ (black), the linear control law obtained from (3.64) using $\rho = 0$ (blue), and the linear feedback controller from [3] (red)

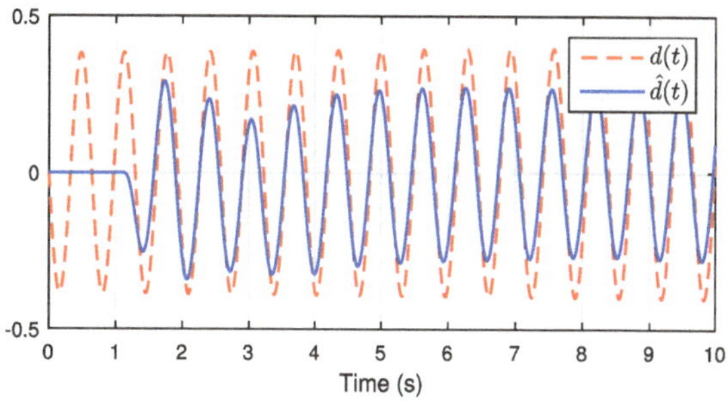

Fig. 3.6 Sinusoidal disturbance signal (dashed) and its estimate (solid), when matrices A, B, and A_d present uncertainties

system. The linear feedback control law proposed in reference [3] (dashed line/red) led to an oscillatory response. The corresponding control signals in this scenario of parametric uncertainties are shown in Fig. 3.5. Figure 3.6 compares the disturbance signal and its estimate obtained by the observer-predictor, when the uncertainties were present, showing an expected estimation error.

This illustrates that the switching term in the proposed control law (3.64) brings robustness with respect to parameter uncertainties for the closed-loop system. They are being compensated since the modulation function $\rho(t)$ of the discontinuous control term satisfies the upper bound (3.65) and adds a degree of freedom to the user/designer in the sense that the parametric uncertainty will be incorporated (at least locally) in the constant η. Hence, the higher η is, more uncertainty is tolerated. Unfortunately, it cannot be guaranteed considering the continuous equivalent control term solely.

3.5.2 *Finite-Time Disturbance Rejection with Prescribed-Time Predictor*

In this subsection, we present the case of constant sensor delays in order to evaluate the delay-compensation strategy proposed in Sect. 3.4 based on the prescribed-time observer (3.89) and the sliding-mode control law (3.64), with controller parameters $k = 2$ and $\rho(t) = 0.5$. It was assumed a constant delay of $D = 1$ seconds. The blow-up function $c(t)$ in (3.77) has parameters $\bar{c}_0 = 2.5$, $T = 9$, and $t_0 = 0$. The prescribed-time observer has time-varying gains $\gamma(t - D)$ given in (3.90), being calculated by using (3.83) and (3.84)–(3.88), with $r_1 = 1$, $r_2 = 3$, $r_3 = 5$.

The successful achievement of the finite-time estimation is shown in Fig. 3.8a–d. As expected, the proposed closed-loop strategy is able to reject completely the effect of the disturbance $d(t)$ in the output signal $y(t)$, see Fig. 3.8e. On the other hand, Fig. 3.8f shows that the sliding variable $\hat{\sigma}(t)$ takes less than 3 seconds to reach the sliding motion in $\hat{\sigma}(t) \equiv 0$. Figure 3.7 compares the disturbance signal $d(t)$ and its estimate $\hat{d}(t)$ obtained by the prescribed-time observer (Fig. 3.8).

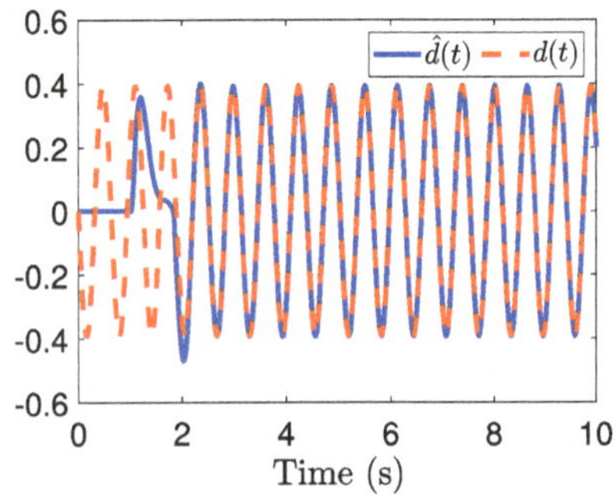

Fig. 3.7 Sinusoidal disturbance signal $d(t)$ (dashed) and its estimate $\hat{d}(t)$ (solid)

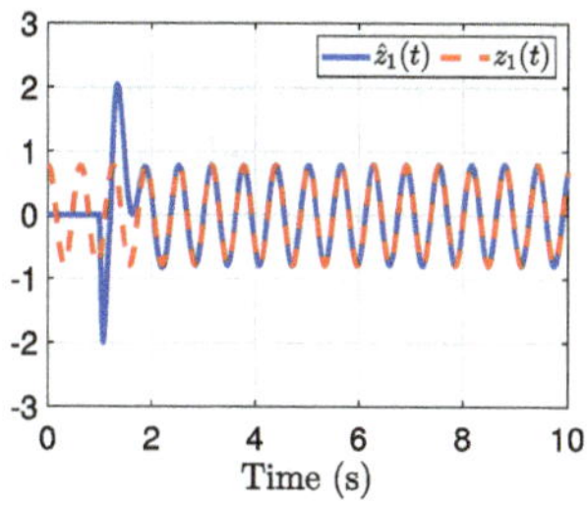

(a) Disturbance component $z_1(t) = x_{d1}(t)$ (dashed) and its estimate $\hat{z}_1(t)$ (solid).

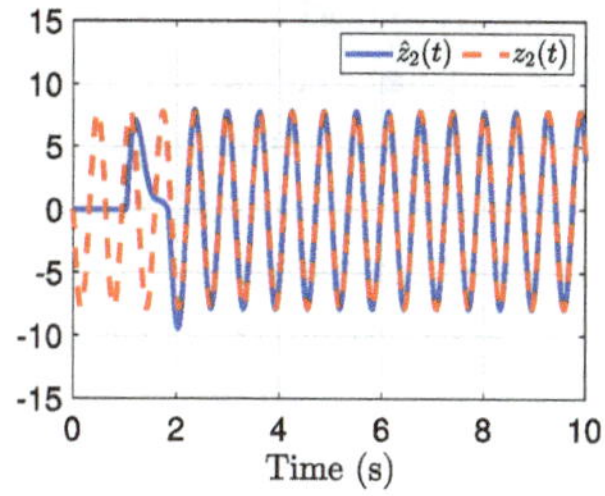

(b) Disturbance component $z_2(t) = x_{d2}(t)$ (dashed) and its estimate $\hat{z}_2(t)$ (solid).

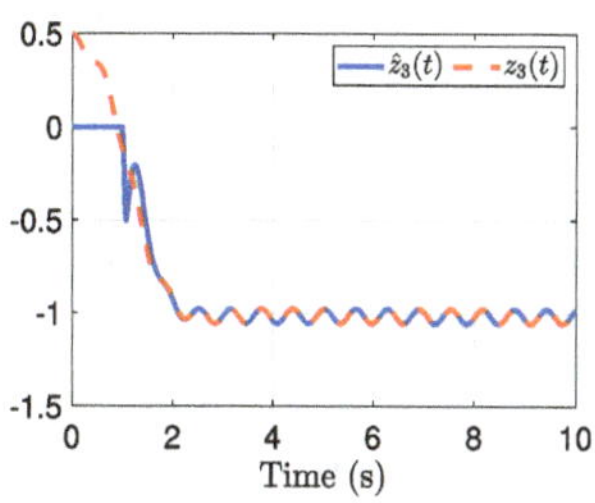

(c) Disturbance component $z_3(t) = x_1(t)$ (dashed) and its estimate $\hat{z}_3(t)$ (solid).

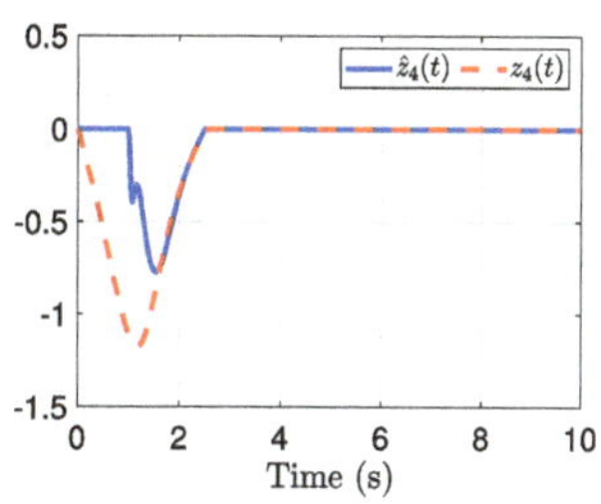

(d) Disturbance component $z_4(t) = x_2(t)$ (dashed) and its estimate $\hat{z}_4(t)$ (solid).

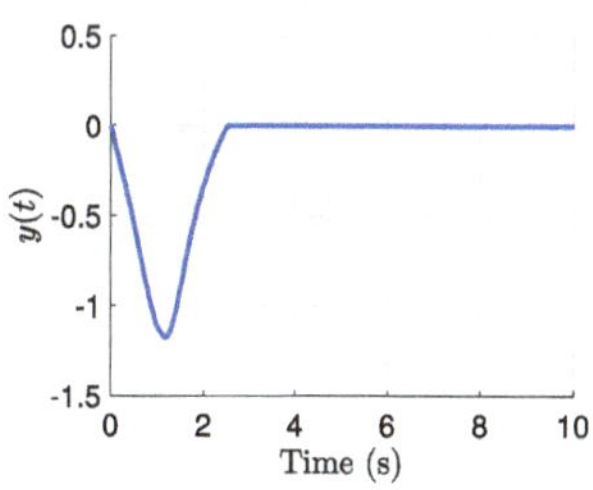

(e) Output signal, $y(t)$.

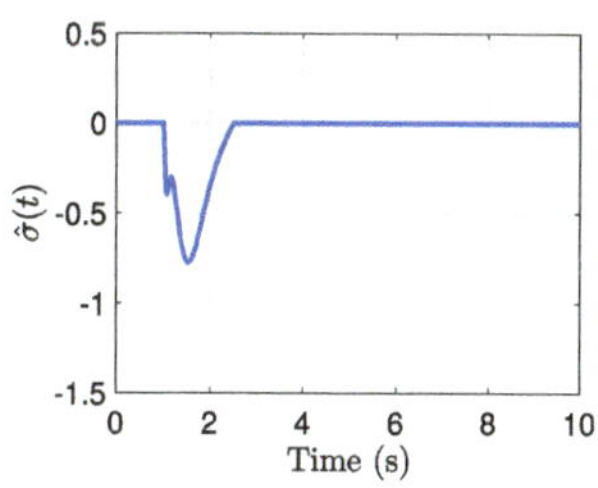

(f) Sliding variable, $\hat{\sigma}(t)$.

Fig. 3.8 Closed-loop response with delay-compensated prescribed-time observer

3.6 Notes and References

As discussed in the preface of this brief, time delays are present on either actuation or sensing paths [21] in most control engineering applications, such as chemical process control, machining, combustion systems, teleoperation, and networked systems, to name only a few. The presence of significant measurement delays in control loops may cause poor performance or instability. For stable linear systems with known constant time delays, a control scheme based on the *Smith predictor* [35] is well-known. In [22, 23], a time-varying input delay is compensated using a predictor approach using the *PDE-backstepping* method [24, 36], by representing the time delay as a transport PDE. More recently, more general classes of distributed unbounded delays are also addressed by the backstepping approach [44].

For systems with parametric uncertainties or unmeasured disturbances, sliding-mode control (SMC) is a well-known and attractive control strategy, due to its robustness, good transient performance, and disturbance rejection capability [8, 13, 41]. It is also a natural choice for systems with switched actuators (such as power converters) [7]. When the state vector is not measurable, the use of state observers, as originally proposed by Bondarev et al. [4], for linear systems with known parameters, is an effective solution, whereby chattering, the foremost flaw of SMC, can be avoided even in the presence of unmodeled dynamics [41].

The in-sensitiveness of SMC to matched disturbances, i.e., which belong to the range of the control distribution matrix, is well-known. The case of unmatched disturbances is more difficult. Examples of SMC for rejection of such disturbances can be found in [14, 45], using a disturbance observer. However, it should be noted that full-state vector is assumed measurable. Other observer designs can simultaneously estimate state and disturbance [30]. A comprehensive overview of such designs can be seen in the survey papers [29] and [5].

Several papers dealt with SMC for systems presenting time delay in the state vector [6, 25, 27, 39]. For systems with measurement delay, there are considerably fewer references. In the case of output delay, Shtessel et al. [33] used a Padé approximation to transform the control problem to that of controlling systems with nonminimum phase. The proposed method was restricted to small constant delays. In contrast, in [7], arbitrarily large delays were dealt with based on a cascade of high-gain predictors for obtaining an asymptotic estimate of the current system state. A SMC law was then used to achieve global stabilization even in the presence of (known) nonlinearities of a given class. The cascade strategy [1] can only be applied to constant measurement delays, and, even for short delay durations, the number of observers in such a cascade may easily reach thousands, which seems highly unpractical [15].

In this chapter, we considered the output regulation problem with disturbance rejection for systems with time-varying measurement delay under the action of an unmeasured disturbance. To this end, we extend the PDE-backstepping approach derived in [22], to design an unknown input observer-predictor for systems with measurement delay. We assume that the disturbance is modeled by an external dynamic system [3]. Then an output feedback SMC law is designed to reject the matched or unmatched disturbances. To the best of our knowledge of the existing literature, this is the first result on the design of a sliding-mode controller (SMC), producing ideal sliding mode and providing global exponential output regulation to zero despite the presence of matched or unmatched disturbance, for arbitrarily long, possibly time-varying, measurement delay. Additionally, in the case of arbitrarily large constant delays, a prescribed-time PDE predictor [11] can be designed in order to obtain finite-time (faster than exponential) disturbance rejection SMC schemes. This inclusion is important since in the previous results only asymptotic results could be proved. Although this kind of observer had already been proposed in reference [11], such method has never been tested in closed loop.

References

1. Ahmed-Ali T, Cherrier E, Lamnabhi-Lagarrigue F (2012) Cascade high gain predictors for a class of nonlinear systems. IEEE Trans Autom Control 57(1):224–229
2. Antsaklis P, Michel A (2006) Linear systems, 1st edn. corr. 2nd printing. Birkhäuser, Boston
3. Åström, KJ, Wittenmark B (2001) Computer-controlled systems. Prentice-Hall, Upper Saddle River
4. Bondarev AG, Bondarev SA, Kostyleva NE, Utkin VI (1985) Sliding modes in systems with asymptotic state observers. Automat Rem Contr 46:679–684
5. Chen WH, Yang J, Guo L, Li S (2016) Disturbance-observer-based control and related methods: an overview. IEEE Trans Ind Electron 63(2):1083–1095
6. Coutinho CL, Oliveira TR, Cunha JPVS (2013) Output-feedback sliding-mode control of multivariable systems with uncertain time-varying state delays and unmatched non-linearities. IET Control Theory Appl 7(12):1616–1623
7. Coutinho CL, Oliveira TR, Cunha JPVS (2014) Output-feedback sliding mode control via cascade observers for global stabilisation of a class of nonlinear systems with output delay. Int J Control 87(11):2327–2337
8. Edwards C, Spurgeon SK (1998) Sliding mode control: theory and applications. Taylor & Francis, London
9. Espitia N, Perruquetti W (2022) Predictor-feedback prescribed-time stabilization of LTI systems with input delay. IEEE Trans Autom Control 67:2784–2799
10. Espitia N, Polyakov A, Efimov D, Perruquetti W (2019) Boundary time-varying feedbacks for fixed-time stabilization of constant-parameter reaction–diffusion systems. Automatica 103:398–407
11. Espitia N, Steeves D, Perruquetti W, Krstic M (2022) Sensor delay-compensated prescribed-time observer for LTI systems. Automatica 135:110005
12. Filippov AF (1964) Differential equations with discontinuous right-hand side. Amer Math Soc Transl 42(2):199–231
13. Fridman L (2019) Special issue: Utkin 80 – sliding mode control and observation. Int J Robust Nonlin Control 29:519–521
14. Ginoya D, Shendge PD, Phadke SB (2014) Sliding mode control for mismatched uncertain systems using an extended disturbance observer. IEEE Trans Ind Electron 61(4):1983–1992
15. Holloway J, Krstic M (2016) Predictor observers for proportional navigation systems subjected to seeker delay. IEEE Trans Contr Syst Technol 24(6):2002–2015
16. Holloway J, Krstic M (2019) Prescribed-time observers for linear systems in observer canonical form. IEEE Trans Autom Control 64(9):3905–3912
17. Jardine A, Lin D, Banjevic D (2006) A review on machinery diagnostics and prognostics implementing condition-based maintenance. Mech Syst Sig Proces 20(7):1483–1510
18. Jimenez-Rodriguez E, Sanchez-Torres JD, Loukianov AG (2017) On optimal predefined-time stabilization. Int J Robust Nonlin Control 27:3620–3642
19. Khalil HK (2002) Nonlinear systems, 3rd edn. Prentice Hall, Hoboken
20. Klamka J (1982) Observer for linear feedback control of systems with distributed delays in controls and outputs. Syst Control Lett 1(5):326–331
21. Krstic M (2010) Compensation of infinite-dimensional actuator and sensor dynamics. IEEE Contr Syst Mag 30(1):22–41
22. Krstic M (2009) Delay compensation for nonlinear, adaptive, and PDE systems. Birkhäuser, Basel
23. Krstic M (2010) Lyapunov stability of linear predictor feedback for time-varying input delay. IEEE Trans Automat Contr 55(2):554–559
24. Krstic M, Smyshlyaev A (2008) Backstepping boundary control for first-order hyperbolic PDEs and application to systems with actuator and sensor delays. Syst Control Lett 57(9):750–758

25. Li X, DeCarlo RA (2003) Robust sliding mode control of uncertain time delay systems. Int J Control 76(13):1296–1305
26. Neerhoff F-L, Van der Kloet P (2003) Canonical representations for single-input single-output linear time-varying systems. In: International workshop on nonlinear dynamics of electronic systems, pp 173–176
27. Orlov Y, Perruquetti W, Richard JP (2003) Sliding mode control synthesis of uncertain time-delay systems. Asian J Control 5(4):568–577
28. Pinto HLCP, Oliveira TR, Hsu L, Krstic M (2018) Sliding mode control for disturbance rejection in systems with measurement delay using PDE-backstepping predictor. In: The 2018 American Control Conference, pp 4099–4104
29. Radke A, Gao Z (2006) A survey of state and disturbance observers for practitioners. In: IEEE 2006 American Control Conference - Minneapolis, MN, pp 5183–5188
30. Rios H, Davila J, Fridman L, Edwards C (2015) Fault detection and isolation for nonlinear systems via high-order-sliding-mode multiple-observer. Int J Robust Nonlin Control 25:2871–2893
31. Seeber R (2023) Generalized Filippov solutions for systems with prescribed-time convergence. CoRR abs/2303.03133, pp 1–7. https://doi.org/10.48550/arXiv.2303.03133
32. Shtessel Y, Edwards C, Fridman L, Levant A (2014) Sliding mode control and observation, 1st edn. Springer, New York
33. Shtessel YB, Zinober ASI, Shkolnikov IA (2003) Sliding mode control for nonlinear systems with output delay via method of stable system center. J Dyn Syst-Trans ASME 125(2):253–257
34. Silverman L-M (1966) Transformation of time-variable systems to canonical (phase-variable) form. IEEE Trans Autom Control 11:300–303
35. Smith OJM (1957) Closer control of loops with dead time. Chem Eng Progr 53(5):217–219
36. Smyshlyaev A, Krstic M (2005) Backstepping observers for a class of parabolic PDEs. Syst Control Lett 54(7):613–625
37. Song Y, Wang Y, Holloway J, Krstic M (2017) Time-varying feedback for regulation of normal-form nonlinear systems in prescribed finite time. Automatica 83:243–251
38. Song Y-D, Wang Y-J, Krstic M (2019) Time-varying feedback for stabilization in prescribed finite time. Int J Robust Nonlin Control 29:618–633
39. Souza AR, Efimov D, Raissi T (2022) Robust output feedback model predictive control of time-delayed systems using interval observers. Int J Robust Nonlin Control 32:1180–1193
40. Tsakalis K, Ioannou P (1993) Linear time-varying systems: control and adaptation. Prentice-Hall, Inc., Hoboken
41. Utkin VI (1992) Sliding modes in control and optimization. Springer, Berlin
42. Utkin V (2016) Discussion aspects of high-order sliding mode control. IEEE Trans Autom Control 61:829–833
43. Watanabe K, Ito M (1981) An observer for linear feedback control laws of multivariable systems with multiple delays in controls and outputs. Syst Control Lett 1(1):54–59
44. Xu X, Liu L, Krstic M, Feng G (2021) Predictor feedback and integrator backstepping of linear systems with distributed unbounded delays. Int J Robust Nonlin Control 1–11. https://doi.org/10.1002/rnc.5810
45. Yang J, Li S, Yu X (2013) Sliding-mode control for systems with mismatched uncertainties via a disturbance observer. IEEE Trans Ind Electron 60(1):160–169

Chapter 4
Disturbance Rejection with PDE-Backstepping Predictor and Super-Twisting Observer

Chapter 3 did not consider the case of nonmodeled disturbances (or unstructured faults), nor did it address the robustness of the proposed methods with respect to parametric uncertainties. In this sense, this chapter presents an advanced control methodology for linear time-invariant uncertain systems affected by known time-varying measurement delays and matched unknown nonlinear disturbances, including actuator faults. Given that certain state variables remain unmeasured at present time, the approach combines an open-loop predictor with a state observer based on the super-twisting algorithm to simultaneously compensate for delays and estimate unmeasured states. Unlike conventional models relying on exogenous dynamic systems, we introduce a nonlinear observer structure that allows the reconstruction of nonmodeled fault signals while remaining robust to parametric uncertainties. Basically, as in the previous chapter, the predictor still advances the delayed output signal in time, ensuring ideal sliding-mode operation. The developed sliding-mode control law achieves asymptotic stability despite perturbations, nonmodeled disturbances, and parameter uncertainties.

4.1 Problem Formulation

Consider a linear time-invariant system subject to actuator faults, which is represented by an unknown input disturbance $d(t) \in \mathbb{R}^p$, as well as parametric uncertainties encompassed in the term $\delta(x, t) \in \mathbb{R}^h$:

$$\begin{cases} \dot{x}(t) = Ax(t) + B\left[u(t) + d(t)\right] + D\delta(x, t)\,, \\ y(t) = Cx(t - \tau(t)), \end{cases} \tag{4.1}$$

H. L. C. P. Pinto et al., *Sliding-Mode Control and Estimation for Time-Delay Systems*, SpringerBriefs in Control, Automation and Robotics,
https://doi.org/10.1007/978-3-032-30178-9_4

where $x(t) \in \mathbb{R}^n$ is the state vector and $u(t) \in \mathbb{R}^m$ is the control input. The output signal $y(t) \in \mathbb{R}^p$, $h \leq p < n$, has a known time-varying measurement delay, denoted by $\tau(t)$. The matrices $A \in \mathbb{R}^{n\times n}$, $B \in \mathbb{R}^{n\times m}$, $D \in \mathbb{R}^{n\times h}$, and $C \in \mathbb{R}^{p\times n}$ are also assumed to be known.

For the sake of simplicity, we adopt the notation $x := x(t)$ for variables in the current time and $x_\tau := x(t - \tau(t))$ for delayed variables.

In this chapter, the state vector x is assumed to be partially unknown such that (4.1) can be rewritten as

$$\begin{cases} \begin{bmatrix} \dot{x}_1 \\ \dot{x}_2 \end{bmatrix} = \begin{bmatrix} A_{11} & A_{12} \\ A_{21} & A_{22} \end{bmatrix} \begin{bmatrix} x_1 \\ x_2 \end{bmatrix} + \begin{bmatrix} B_1 \\ 0 \end{bmatrix} (u + d) \\ \qquad + \begin{bmatrix} D_1 \\ D_2 \end{bmatrix} \delta(x, t), \\ y(t) \quad = x_{\tau,2}, \end{cases} \tag{4.2}$$

with $x_{\tau,2} := x_2(t - \tau(t))$, $x_1 \in \mathbb{R}^{n-p}$ is measured, and $x_2 \in \mathbb{R}^p$ is unmeasured. The matrices A_{11}, A_{12}, A_{21}, A_{22}, and B_1 have appropriate dimensions. For instance, in this class of systems, x_1 may represent the state of an actuator dynamics, such as pilot valves in oil industry applications, while x_2 governs the process dynamics, whose effects will be realized just after some delay in the output signal.

The following assumptions are further assumed for the system (4.2):

(A1) The state vector $x_1(t)$ is assumed to be measured in the current time.
(A2) The actuator fault $d(t)$ is uniformly bounded such that:

$$\|d(t)\| \leq \alpha. \tag{4.3}$$

(A3) The rate of variation of the delay $\dot{\tau}$ is norm bounded by some known constant $0 < \bar{r} < 1$ such that

$$\|\dot{\tau}\| \leq \bar{r}. \tag{4.4}$$

(A4) The pair (A, B) is controllable.
(A5) The matrix B has full rank.
(A6) The effects of the parametric uncertainties $\delta(x, t)$ are norm bounded by

$$\|\delta(x, t)\| \leq \bar{\delta}\|x\|, \tag{4.5}$$

where $\bar{\delta} > 0$ is a known constant. The objective of this chapter is to propose a control law that allows us to stabilize the time-delay system even in the presence of actuator faults $d(t)$ and parametric uncertainties. Moreover, we pursue the fault signal reconstruction for diagnosis purposes.

4.2 Observer and Predictor

In this section, an observer-predictor scheme is developed that is capable of estimating in current time the state variable x_2 as well as estimating (reconstructing) the disturbance signal (actuator fault) $d(t)$ from the measured output $y(t)$.

4.2.1 *Predictor for Measurement Delays*

Initially, for the sake of clarity, consider the system (4.2) without parametric uncertainties ($\delta(x,t) = 0$). Assuming **(A1)**, it is possible to verify [22] that the x_2-dynamics can be seen as a linear system with a known input x_1 and the output $y = x_2(t - \tau(t))$. Hence, it is not difficult to obtain an estimate $\hat{x}_2$ from the exact solution of the system without parametric uncertainties via the variation of constants formula [2, 14, 16], given by

$$\hat{x}_2 = e^{A_{22}\tau(t)}x_{\tau,2} + \int_{t-\tau(t)}^{t} e^{A_{22}(t-\theta)}A_{21}x_1(\theta)d\theta. \tag{4.6}$$

The main advantage of this approach is that if the system is totally known, the estimate $\hat{x}_2$ is an exact prediction of x_2 obtained in finite time from its delayed value $x_{\tau,2}$.

In the presence of parametric uncertainties, the exact solution of the x_2-dynamics would be given by

$$\begin{aligned} x_2 &= e^{A_{22}\tau(t)}x_{\tau,2} \\ &+ \int_{t-\tau(t)}^{t} e^{A_{22}(t-\theta)}\left[A_{21}x_1(\theta) + D_2\delta(x,\theta)\right]d\theta. \end{aligned} \tag{4.7}$$

In this sense, using the predictor (4.6), the prediction would be affected by a residual error $\tilde{x}_2 = x_2 - \hat{x}_2$. On the other hand, from (4.6) and (4.7), we can verify that

$$\tilde{x}_2 = \int_{t-\tau(t)}^{t} e^{A_{22}(t-\theta)}D_2\delta(x,\theta)d\theta. \tag{4.8}$$

From this expression (4.8), we can note that the delay duration has a significant influence on the magnitude of such an error, which can be majorized by

$$\|\tilde{x}_2\| \leq \tau(t)\bar{\delta}\|e^{A_{22}\tau(t)}D_2\| \sup_{\theta\in[t-\tau(t),t]} (\|x(\theta)\|). \tag{4.9}$$

4.2.2 Nonlinear Super-Twisting Observer

From the estimate $\hat{x}_2$, it is possible to employ a second-order SMO using the *super-twisting algorithm* (STA) [6, 20] in order to obtain an estimate of the actuator fault $d(t)$. By defining the estimation error:

$$\tilde{x}_1 = x_1 - \hat{x}_1, \tag{4.10}$$

we can design the following nonlinear observer:

$$\begin{cases} \dot{\hat{x}}_1 &= A_{11}\hat{x}_1 + A_{12}\hat{x}_2 + B_1 u + \nu \\ \nu &= -k_1 \frac{\tilde{x}_1}{\|\tilde{x}_1\|^{1/2}} + k_2 \tilde{x}_1 + \hat{\xi} \\ \dot{\hat{\xi}} &= -k_3 \frac{\tilde{x}_1}{\|\tilde{x}_1\|} - k_4 \tilde{x}_1 \,, \end{cases} \tag{4.11}$$

where $\hat{x}_1$ is the estimate of the state vector x_1 and $\hat{x}_2$ is the predicted value of x_2, obtained from the proposed observer-predictor scheme of Sect. 4.2.1. Notice that the prediction $\hat{x}_2$ obtained before is exact in the case where there are no uncertainties in plant parameters or delay. In the case where uncertainties are present, the residual error $\tilde{x}_2$ is not necessarily zero. Thus, the observer based on the STA is able to handle the uncertainties as part of the fault signal.

From assumptions **(A2)** and **(A6)**, we can directly apply the procedure of [20, eq. (25)] in order to determine the gains $k_1, \cdots, k_4$ and guarantee that the sliding mode $\dot{\tilde{x}}_1 = \tilde{x}_1 \equiv 0$ is achieved in finite time. Similar steps to those presented in [17] and [19] can also be employed to obtain such gains.

The dynamic equation for the estimation error $\tilde{x}_1$ can be written by including the effects of the uncertainties as follows:

$$\begin{cases} \dot{\tilde{x}}_1 &= A_{11}\tilde{x}_1 + A_{12}\tilde{x}_2 + B_1 d + D_1\delta(x,t) - \nu \\ \nu &= -k_1 \frac{\tilde{x}_1}{\|\tilde{x}_1\|^{1/2}} + k_2 \tilde{x}_1 + \hat{\xi} \\ \dot{\hat{\xi}} &= -k_3 \frac{\tilde{x}_1}{\|\tilde{x}_1\|} - k_4 \tilde{x}_1 \,. \end{cases} \tag{4.12}$$

After the sliding mode is reached, we can write

$$\begin{cases} 0 = 0 + A_{12}\tilde{x}_2 + B_1 d + D_1\delta(x,t) - \nu \\ \nu = \hat{\xi}, \quad \dot{\hat{\xi}} = 0 \,. \end{cases} \tag{4.13}$$

Rearranging the first equation in (4.13), we obtain

$$\hat{\xi} = B_1 d + D_1\delta(x,t) + A_{12}\tilde{x}_2, \tag{4.14}$$

which represents an estimate of the fault and disturbances effects in x_1. Notice that in the uncertainties-free case, the signal $\hat{\xi}$ can be directly used to obtain the fault reconstruction of $d(t)$.

4.2.3 Observer-Predictor Combination

Now, we can combine the predictor (4.6) and the observer (4.11) in a single observer-predictor scheme for the uncertain system with output delays and partially unmeasured states:

$$\begin{cases} \hat{x}_2 &= e^{A_{22}\tau(t)} x_{\tau,2} \\ &\quad + \int_{t-\tau(t)}^{t} e^{A_{22}(t-\theta)} A_{21} x_1(\theta) d\theta \\ \dot{\hat{x}}_1 &= A_{11}\hat{x}_1 + A_{12}\hat{x}_2 + B_1 u + \nu \\ \nu &= -k_1 \frac{\tilde{x}_1}{\|\tilde{x}_1\|^{1/2}} + k_2 \tilde{x}_1 + \hat{\xi} \\ \dot{\hat{\xi}} &= -k_3 \frac{\tilde{x}_1}{\|\tilde{x}_1\|} - k_4 \tilde{x}_1 \,, \end{cases} \tag{4.15}$$

where $\tilde{x}_1 = x_1 - \hat{x}_1$. Such a unified observer-predictor scheme would be able to ultimately estimate (even in finite time) the state x_2 as well as the disturbance term aggregated with parametric uncertainties as the state vanishes—see **(A6)**. Again, in the uncertainty-free case, we can also reconstruct the actuator fault $d(t)$ directly from the signal $\hat{\xi}$ by means of (4.14), according to assumption **(A5)**.

4.3 Sliding-Mode Control

Once the estimates $\hat{x}_2$ and $\hat{\xi}$ are obtained, from assumption **(A4)**, we can design an SMC law in order to obtain the asymptotic stabilization result along with robustness to model uncertainties and perturbations.

4.3.1 Control Law

First of all, let us define the following state vector:

$$\bar{x} = \begin{bmatrix} x_1 \\ \hat{x}_2 \end{bmatrix}, \tag{4.16}$$

which contains the measured state x_1 and the estimate $\hat{x}_2$, obtained by means of the predictor (4.6), both of them in the current time.

Now, let us define the sliding variable

$$s(t) = S\bar{x} = \begin{bmatrix} I & S_2 \end{bmatrix} \begin{bmatrix} x_1 \\ \hat{x}_2 \end{bmatrix}, \tag{4.17}$$

where S is an appropriate matrix to be designed such that SB is not singular.

From (4.6), we get

$$\begin{aligned}\dot{\hat{x}}_2 &= (1-\dot{\tau})e^{A_{22}\tau}\dot{x}_{\tau,2} + \dot{\tau}A_{22}e^{A_{22}}x_{\tau,2} \\ &+ A_{21}x_1 - (1-\dot{\tau})e^{A_{22}\tau}A_{21}x_{\tau,1} \\ &+ A_{22}\int_{t-\tau}^{t} e^{A_{22}(t-\theta)}A_{21}x_1(\theta)d\theta. \end{aligned} \tag{4.18}$$

Using (4.2) and (4.6) in order to perform appropriate substitutions, we can check that

$$\dot{\hat{x}}_2 = A_{21}x_1 + A_{22}\hat{x}_2 + (1-\dot{\tau})e^{A_{22}\tau}D_2\delta(x, t-\tau). \tag{4.19}$$

In order to make the notation simpler, let us define

$$\zeta_1(x,t) := D_1\delta(x,t), \tag{4.20}$$

$$\zeta_2(x,t) := (1-\dot{\tau})e^{A_{22}\tau}D_2\delta(x, t-\tau). \tag{4.21}$$

Now, consider the linear transformation

$$T := \begin{bmatrix} I & S_2 \\ 0 & I \end{bmatrix}, \tag{4.22}$$

such that

$$\begin{bmatrix} s \\ \hat{x}_2 \end{bmatrix} = T\begin{bmatrix} x_1 \\ \hat{x}_2 \end{bmatrix}. \tag{4.23}$$

Thus, we can write in the new coordinates the following dynamic system:

$$\begin{cases} \begin{bmatrix} \dot{s} \\ \dot{\hat{x}}_2 \end{bmatrix} = \underbrace{\begin{bmatrix} \bar{A}_{11} & \bar{A}_{12} \\ \bar{A}_{21} & \bar{A}_{22} \end{bmatrix}}_{\bar{A}}\begin{bmatrix} s \\ \hat{x}_2 \end{bmatrix} + \begin{bmatrix} B_1 \\ 0 \end{bmatrix}(u+d) \\ \qquad + \begin{bmatrix} \zeta_1(x,t) + S_2\zeta_2(x,t) \\ \zeta_2(x,t) \end{bmatrix}, \\ y(t) = x_{\tau,2}, \end{cases} \tag{4.24}$$

where $\bar{A}_{11} = A_{11} + S_2A_{21}$, $\bar{A}_{12} = S_2A_{22} + A_{12} - \bar{A}_{11}S_2$, $\bar{A}_{21} = A_{21}$, and $\bar{A}_{22} = A_{22} - A_{21}S_2$.

In order to design the matrix S, let us initially consider equation (4.24) for $\dot{\hat{x}}_2$, free of parametric uncertainties, i.e., when $\zeta_2(x,t) = 0$. During the sliding motion, $s = 0$ and

$$\dot{\hat{x}}_2 = \bar{A}_{22}\hat{x}_2 \,. \tag{4.25}$$

Hence, during the sliding mode, the $\hat{x}_2$-dynamics is governed by the eigenvalues of $\bar{A}_{22} = A_{22} - A_{21}S_2$. On the other hand, assumption **(A4)** assures that the pair (A_{22}, A_{21}) is controllable (see [8, Prop. 3.3]), and, thus, any pole-placement technique can be employed in order to obtain S_2 such that $\bar{A}_{22}$ is Hurwitz.

Consider the control law

$$u = u_d + u_{\text{nom}} + u_{\text{SM}}, \tag{4.26}$$

where u_d is the term designed for disturbance rejection, whereas u_{nom} is the nominal part of the control signal and u_{SM} is the switching term that serves in order to drive the sliding variable s to the sliding mode $s \equiv 0$.

Note that the x_1-dynamics in (4.2) is

$$\dot{x}_1 = A_{11}x_1 + A_{12}x_2 + B_1(u + d) + \zeta_1(x,t). \tag{4.27}$$

From (4.14), we can define

$$u_d = -B_1^{\dagger}\hat{\xi} = -d - B_1^{\dagger}D_1\delta(x,t) - B_1^{\dagger}A_{12}\tilde{x}_2, \tag{4.28}$$

where $B_1^{\dagger}$ is the Moore–Penrose pseudo-inverse of B_1, given by $B_1^{\dagger} = (B_1^T B_1)^{-1} B_1^T$ [26, page 122].

Plugging (4.28) into (4.27), we can verify that

$$\dot{x}_1 = A_{11}x_1 + A_{12}\hat{x}_2 + B_1[u_{\text{nom}} + u_{\text{SM}}], \tag{4.29}$$

since $\tilde{x}_2 = x_2 - \hat{x}_2$. Notice that u_d is able to reject the disturbance d as well as the matched part of the parametric uncertainties $A_{12}\tilde{x}_2$ and $\zeta_1(x,t)$.

Considering the control law u_d, we can design u_{nom} by means of the equivalent control method [23] along with the solution of the equation $\dot{s} = 0$ for the nominal system (without uncertainties), and taking into account that the x_1-dynamics is already in the form (4.29). Thus, supposing $u_{\text{SM}} = 0$, we can write

$$0 = SA\bar{x} + SB(u_{\text{nom}}). \tag{4.30}$$

Since SB is not singular by construction, we have

$$u_{\text{nom}} = -(SB)^{-1}SA\bar{x}. \tag{4.31}$$

Finally, we define the switching term to drive the closed-loop system to a sliding mode:

$$u_{\text{SM}} = -\rho(SB)^{-1}\text{sign}(s), \tag{4.32}$$

where $\text{sign}(s) = \left[\text{sign}(s_1) \ \cdots \ \text{sign}(s_{n-p})\right]^T$ and $\rho > 0$ must be chosen sufficiently large in order to majorize the uncertainties and disturbances not compensated by the other control terms u_d and u_{nom}.

Hence, the complete expression of the control law can be written as follows:

$$u = -B_1^{\dagger}\hat{\xi} - (SB)^{-1}SA\bar{x} - \rho(SB)^{-1}\text{sign}(s). \tag{4.33}$$

4.3.2 *Stability Analysis*

By plugging (4.33) into (4.24), we get

$$\begin{cases} \dot{s} &= S_2\zeta_2(x,t) - \rho\text{sign}(s) \\ \dot{\hat{x}}_2 &= \bar{A}_{21}s + \bar{A}_{22}\hat{x}_2 + \zeta_2(x,t) \\ y &= x_{\tau,2} = \hat{x}_{\tau,2} + \tilde{x}_{\tau,2}. \end{cases} \tag{4.34}$$

From the closed-loop system (4.34), it is possible to verify that the sliding mode $s \equiv 0$ is achieved in finite time, according to the next theorem.

Theorem 4.1 *The control law (4.33) applied to (4.2), under the assumptions* ***(A1)*** *to* ***(A6)****, with*

$$\rho \geq \phi\bar{\delta}(1+\bar{r})\|e^{A_{22}\tau}D_2\|\|x\| + \eta\,, \tag{4.35}$$

$\eta > 0$ *and* $\phi > 1$ *being arbitrary constants, guarantees the existence of the sliding mode* $s = S\bar{x} \equiv 0$ *in finite time.*

Proof Consider the Lyapunov function $V = s^Ts/2$. Calculating the time derivative of V along the trajectories of (4.34) and using the definition of $\zeta_2(x,t)$ in (4.21), we obtain

$$\dot{V} = s^TS_2\zeta_2(x,t) - \rho s^T\text{sign}(s), \tag{4.36}$$

$$\dot{V} = (1-\dot{\tau})s^Te^{A_{22}\tau}D_2\delta(x,t-\tau) - \rho s^T\text{sign}(s). \tag{4.37}$$

Notice that

$$s^T\text{sign}(s) = \|s\|_1 \geq \|s\|_2, \tag{4.38}$$

$$-\rho s^T\text{sign}(s) \leq -\rho\|s\|_2. \tag{4.39}$$

Since $\dot{V}$ contains a delayed term, we will employ the Razumikhin's theorem [10, Theorem 1.4] in the following developments. First, reminding that $\|x(t+\theta)\| < \phi\|x(t)\|$, for $\phi > 1$, $-\tau(t) \le \theta \le 0$. Thus,

$$\|\delta(x, t+\theta)\| \le \bar{\delta}\|x(t+\theta)\| \le \phi\bar{\delta}\|x(t)\|, \tag{4.40}$$

$$\|\delta(x, t-\tau)\| \le \phi\bar{\delta}\|x(t)\|. \tag{4.41}$$

Hence, using **(A3)**, we have

$$\dot{V} \le -\|s\|\left[\rho - \phi\bar{\delta}(1+\bar{r})\|e^{A_{22}\tau}D_2\|\|x(t)\|\right]. \tag{4.42}$$

If

$$\rho \ge \phi\bar{\delta}(1+\bar{r})\|e^{A_{22}\tau}D_2\|\|x\| + \eta, \tag{4.43}$$

then

$$\dot{V} \le -\eta\|s\| < 0, \tag{4.44}$$

showing that s is globally asymptotically stable. Moreover, since

$$\dot{V} \le -\eta\sqrt{V}, \tag{4.45}$$

we conclude that the sliding surface $s \equiv 0$ is achieved in finite time. □

We can note the term dependent on the variation rate of the delay $\dot{\tau}$ in (4.37), which comes from the perturbation signal $\zeta_2(x,t)$. In order to obtain an upper bound presented in (4.42), the term $\dot{\tau}$ must be bounded, which is verified according to assumption **(A3)**. In the case with no parametric uncertainties, we would have $\delta(x,t) = 0$ in (4.37), and such an assumption could be relaxed since $\dot{V}$ would not contain the term dependent on the delay.

Once the sliding mode is achieved, we can verify that the closed-loop system (4.34) is asymptotically stable via the following corollary.

Corollary 4.1 *Under the same assumptions* ***(A1)*** *to* ***(A6)*** *of Theorem 4.1, when the sliding mode is achieved, the closed-loop system (4.34) is asymptotically stable if there exists a matrix* $P_2 > 0$ *such that* $P_2\bar{A}_{22} + \bar{A}_{22}^T P_2 = -I$ *and*

$$\bar{\delta} < \frac{1}{2\phi\lambda_{\max}(P_2)(1+\bar{r})\left(\sqrt{1+\|S_2\|^2}\right)\|e^{A_{22}\tau}D_2\|}, \tag{4.46}$$

where $\phi > 1$ *is an arbitrary constant and* $\lambda_{\max}(P_2)$ *is the largest eigenvalue of the matrix* P_2.

Proof When the sliding mode is achieved, we have $s \equiv 0$, and the resulting system dynamics, obtained from (4.34), is given by

$$\dot{\hat{x}}_2 = \bar{A}_{22}\hat{x}_2 + \zeta_2(x,t). \tag{4.47}$$

Reminding that $\bar{A}_{22}$ is Hurwitz by design, thus we can find a symmetric positive definite matrix $P_2 > 0$ such that $P_2\bar{A}_{22}+\bar{A}_{22}^T P_2 = -I$. Now, consider the Lyapunov function $V_2(\hat{x}_2) = \hat{x}_2^T P_2\hat{x}_2$. Its time derivative along the trajectories of (4.47) is

$$\begin{aligned}\dot{V}_2(\hat{x}_2) &= -\hat{x}_2^T\hat{x}_2 + 2\hat{x}_2^T P_2\zeta_2(x,t) \\ &\leq -\|\hat{x}_2\|^2 + 2\lambda_{\max}(P_2)\|\hat{x}_2\|\|\zeta_2(x,t)\| \\ &= -\lambda_{\max}(P_2)\|\hat{x}_2\|\left[\mu\|\hat{x}_2\| - 2\|\zeta_2(x,t)\|\right],\end{aligned} \tag{4.48}$$

where $\mu := 1/\lambda_{\max}(P_2)$.

From (4.21), we know that

$$\zeta_2(x,t) = (1-\dot{\tau})e^{A_{22}\tau}D_2\delta(x,t-\tau) \tag{4.49}$$

depends on a delayed variable, and thus, we will utilize again the Razumikhin's theorem. During the sliding mode, one has that $x_1 = -S_2\hat{x}_2$, such that $\|x\| \leq \left(\sqrt{1+\|S_2\|^2}\right)\|\hat{x}_2\|$. Thus, from (4.41), we can write

$$\|\delta(x,t-\tau)\| \leq \phi\bar{\delta}\left(\sqrt{1+\|S_2\|^2}\right)\|\hat{x}_2\|. \tag{4.50}$$

In this sense, substituting (4.50) into (4.49), we get

$$\dot{V}_2(\hat{x}_2) \leq -\lambda_{\max}(P_2)\|\hat{x}_2\|^2\left[\mu - \beta_1\right], \tag{4.51}$$

$$\beta_1 := 2\phi\bar{\delta}(1+\bar{r})\left(\sqrt{1+\|S_2\|^2}\right)\|e^{A_{22}\tau}D_2\|. \tag{4.52}$$

If $\beta_1 < \mu$, one has $\dot{V}_2(\hat{x}_2) < 0$, under the conditions of the Razumikhin's theorem, which shows that the $\hat{x}_2$-dynamics is asymptotically stable whenever

$$2\phi\bar{\delta}(1+\bar{r})\left(\sqrt{1+\|S_2\|^2}\right)\|e^{A_{22}\tau}D_2\| < \frac{1}{\lambda_{\max}(P_2)}, \tag{4.53}$$

leading to the condition (4.46). □

From Corollary 4.1, we can see that $\hat{x}_2 \to 0$ as $t \to \infty$. Since in the sliding mode we have $x_1 = -S_2\hat{x}_2$, we can also verify that $x_1 \to 0$. Moreover, $\|x\| \leq \left(\sqrt{1+\|S_2\|^2}\right)\|\hat{x}_2\|$, and, from **(A6)**, we have that the perturbation effects are also driven to zero and, consequently, $\tilde{x}_2 \to 0$ from Eq. (4.8). Thus, we also conclude that $x_2 \to 0$, which demonstrates that the closed-loop system is stabilized in spite of the presence of the actuator faults and the parametric uncertainties, satisfying the condition (4.46).

Notice that the assumption **(A6)** allows us to stabilize the closed loop at the origin. However, it is possible to relax such an assumption by using a more general upper bound for the parametric uncertainties, such as $\|\delta(x,t)\| \leq \bar{\delta}_1\|x\| + \bar{\delta}_2$. In this case, the state variable would be ultimately bounded inside a sufficiently small neighborhood of the origin, rather than converging exactly to the origin.

4.4 Simulations

In order to illustrate the performance of the proposed method, we have initially considered the system (4.1)–(4.2) free of parametric uncertainties $\delta(x,t) = 0$, with

$$A = \begin{bmatrix} -1 & 1 \\ -3 & 1 \end{bmatrix}, \quad B = \begin{bmatrix} 1 \\ 0 \end{bmatrix}, \quad C = \begin{bmatrix} 1 & 0 \end{bmatrix}. \tag{4.54}$$

The actuator fault was chosen as $d(t) = 0.1\sin(2t) + 0.2\cos(3t)$, and a time-varying delay was considered in the output signal $\tau(t) = 0.4 + 0.1\sin(t)s$. The observer gains were determined according to the procedure established in reference [20, eq. (25)], such that $k_1 = k_3 = 5$ and $k_2 = k_4 = 2$. On the other hand, the control law (4.33) was implemented with $S = [1 \ -5]$ and $\rho = 2$, where the control gain ρ was assumed to be constant for the sake of simplicity.

Figure 4.1 shows the state variables x_1 (in blue) and x_2 (in black) as well as the output signal y (in red). The plant is stabilized with the proposed controller using the measured state x_1 as well as the estimates $\hat{\xi}$ and $\hat{x}_2$ obtained from the observer-predictor scheme (4.15). Figure 4.2 compares x_2 (in blue) and its estimate $\hat{x}_2$ (in red), provided by the predictor (4.6), highlighting that the prediction action occurs in finite time. Figure 4.3 shows the actuator fault d and its estimate $\hat{d} = B_1^{\dagger}\hat{\xi}$ obtained from the observer (4.11), which also converges in finite time, reminding

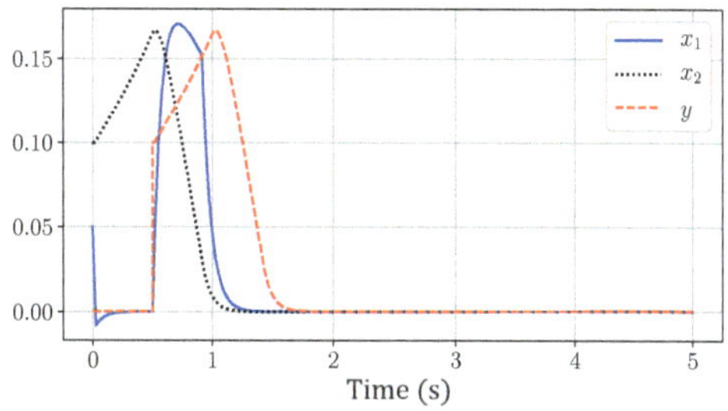

Fig. 4.1 *Without uncertainties:* state variables $x_1(t)$ (in blue), $x_2(t)$ (in black), and the output signal $y(t)$ (in red)

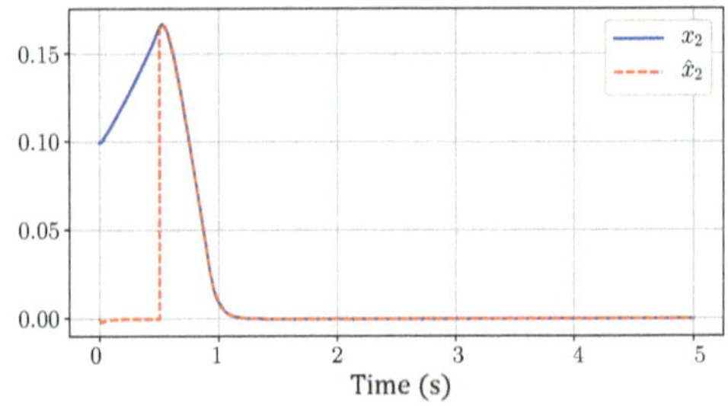

Fig. 4.2 *Without uncertainties:* state variable $x_2(t)$ (in blue) and its estimate $\hat{x}_2(t)$ (in red), obtained with the finite-time predictor (4.6)

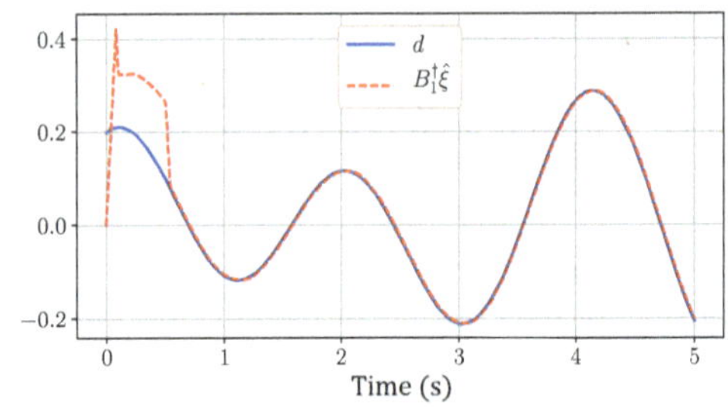

Fig. 4.3 *Without uncertainties:* disturbance $d(t)$ (in blue) and its estimate $\hat{d}(t) = B_1^\dagger \hat{\xi}(t)$ (in red), obtained with the finite-time observer (4.11)

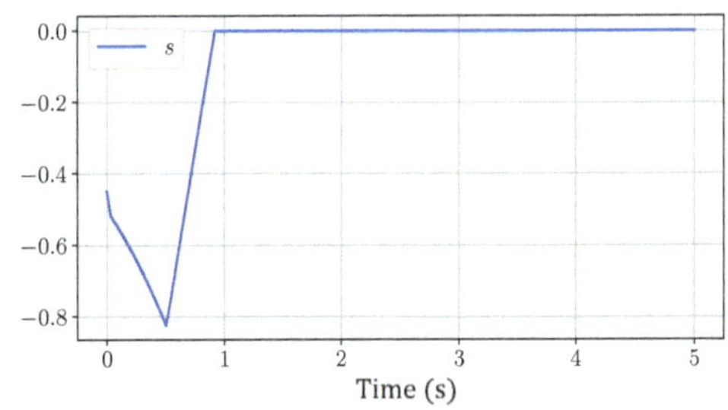

Fig. 4.4 *Without uncertainties:* sliding variable $s(t)$ reaching zero in finite time

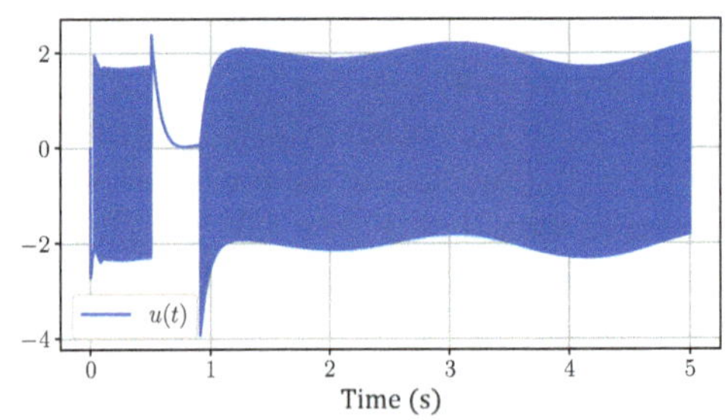

Fig. 4.5 *Without uncertainties:* control signal $u(t)$ highlighting the occurrence of the sliding mode

that $B_1^\dagger = (B_1^T B_1)^{-1} B_1^T$ is the left pseudo-inverse of B_1 [26, page 122]. In Figs. 4.4 and 4.5, we can check the sliding variable s and the control signal u, respectively.

In order to evaluate the robustness of the proposed method in the presence of parametric uncertainties, we have chosen

$$D = \begin{bmatrix} 0.4 & 0.4 \\ 0.4 & 0.4 \end{bmatrix}, \quad \delta(x,t) = x. \tag{4.55}$$

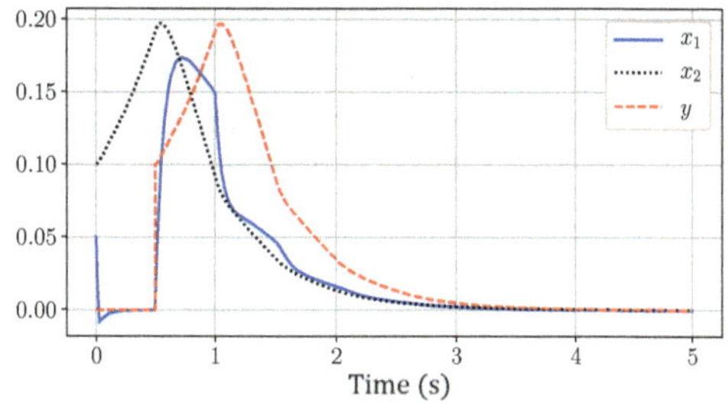

Fig. 4.6 *With uncertainties:* state variables $x_1(t)$ (in blue), $x_2(t)$ (in black), and the output signal $y(t)$ (in red)

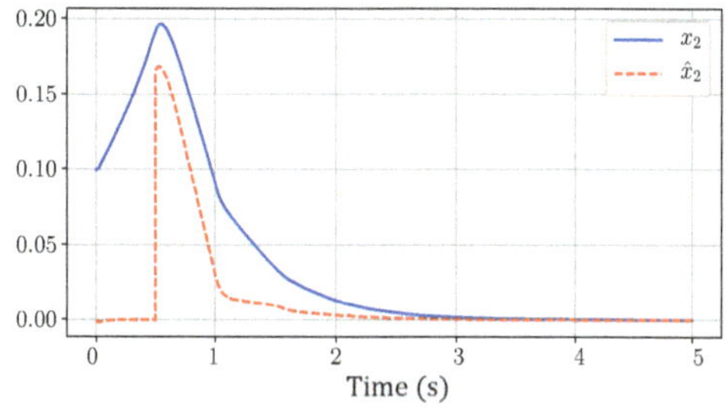

Fig. 4.7 *With uncertainties:* state variable $x_2(t)$ (in blue) and its estimate $\hat{x}_2(t)$ (in red), obtained with the finite-time predictor (4.6)

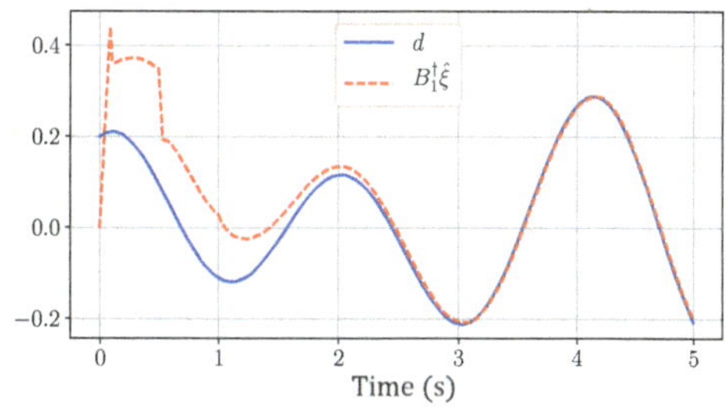

Fig. 4.8 *With uncertainties:* disturbance $d(t)$ (in blue) and its estimate $\hat{d}(t) = B_1^\dagger \hat{\xi}(t)$ (in red), obtained by means of (4.14). The estimate tries to accommodate the uncertainties effects in the prediction error $\tilde{x}_2(t)$

In this second case, the control gain was redesigned to $\rho = 5$ in order to counteract the parametric uncertainties and guarantee the closed-loop stability; otherwise, the system trajectories go unstable (curves not shown). Figure 4.6 shows that the closed-loop system is still stabilized in spite of the parametric uncertainties. Figure 4.7 displays the effects of the parametric uncertainties in the estimation of $\hat{x}_2$. Figure 4.8 illustrates that the nonlinear super-twisting observer reconstructs the disturbance d by ultimately compensating the adverse impacts of the included uncertainties.

4.5 Notes and References

For systems with input-output delays, the control law is usually designed based on some prediction feedback scheme [18]. For stable systems with constant measurement delays, control laws based on the *Smith predictor* [24] are often adopted. This method is also extended to unstable systems or modeled by state-space equations [2]. In [14], time-varying delays of arbitrary duration are compensated using a predictor-based approach derived from the *PDE-backstepping* technique [25], where the delay is represented by a partial differential equation (PDE).

A similar predictor is developed in [16] by means of Lyapunov methods that are able to handle simultaneously input and output delays. Despite being widely used, most of the proposed predictors require the full knowledge of the model and the delay, not being robust to uncertainties or disturbances [15]. The use of adaptive methods allows for circumventing some of these problems [4, 13]. Another strategy widely used for measurement delays is the use of cascade observers [1, 7]. This strategy can be applied to systems with arbitrary measurement delays, but originally they were assumed to be constant [1]. Even in the case of short-duration delays, the number of observers in the cascade may be very large, not facilitating its use in certain applications [11].

The use of remotely driven systems has also contributed to the growing interest in fault reconstruction methods to allow both diagnosis and closed-loop control by estimating the state and disturbances that represent the effects of the faults in the actuators. Among the employed techniques for the simultaneous estimation of the state vector and unknown disturbances, we can highlight unknown input observers (UIO) [5, 12, 21] and sliding-mode observers (SMO) [9, 27]. In [22], a UIO was developed for structured failures, using the *PDE-backstepping* methodology [25], and a sliding-mode control (SMC) law [28] was applied to reject the effect of the actuator fault in the output signal, which was also corrupted by a time-varying delay. The SMC strategy proposed in [22] was based on a structured disturbance observer, where the restrict class of disturbances were supposed to be generated by a linear exogenous (known) dynamic system and no parametric uncertainties were allowed either.

This chapter presented a method to stabilize a linear time-invariant system with partially unavailable states and a known time-varying measurement delay. Unlike earlier approaches in this brief, the proposed methodology enables both nonlinear disturbance rejection and simultaneous reconstruction of unmodeled disturbances, interpreted as fault signals for diagnosis. By combining an open-loop predictor with a second-order sliding-mode observer—the *super-twisting algorithm* (STA) [20]—the full state and actuator fault signals are estimated in real time. A sliding-mode control law then ensures global stabilization despite parametric uncertainties and unmodeled nonlinear faults.

Our results also show that when the predictors compensate for output delays and sliding motion is reached in finite time, asymptotic stabilization, disturbance rejection, and fault reconstruction can be achieved through the proposed sliding-mode controller. Although explicit predictors for uncertain nonlinear time-delay systems are uncommon [3], combining prediction feedback for uncertain linear systems with robust nonlinear control laws broadens the class of time-delay systems that can be handled.

References

1. Ahmed-Ali T, Cherrier E, Lamnabhi-Lagarrigue F (2012) Cascade high gain predictors for a class of nonlinear systems. IEEE Trans Autom Control 57(1):224–229
2. Artstein Z (1982) Linear systems with delayed controls: a reduction. IEEE Trans Autom Control 27(4):869–879
3. Bekiaris-Liberis N, Krstic M (2013) Compensation of state-dependent input delay for nonlinear systems. IEEE Trans Autom Control 58(2):275–289
4. Bresch-Pietri D, Krstic M (2009) Adaptive trajectory tracking despite unknown input delay and plant parameters. Automatica 45(9):2074–2081
5. Chakrabarty A, Corless MJ, Buzzard GT, Zak SH, Rundell AE (2017) State and unknown input observers for nonlinear systems with bounded exogenous inputs. IEEE Trans Autom Control 62(11):5497–5510
6. Chalanga A, Kamal S, Fridman L, Bandyopadhyay B, Moreno JA (2016) Implementation of super-twisting control: super-twisting and higher order sliding-mode observer-based approaches. IEEE Trans Ind Electron 63(6):3677–3685
7. Coutinho CL, Oliveira TR, Cunha JPVS (2014) Output-feedback sliding mode control via cascade observers for global stabilisation of a class of nonlinear systems with output delay. Int J Control 87(11):2327–2337
8. Edwards C, Spurgeon SK (1998) Sliding mode control: theory and applications. Taylor & Francis, London
9. Edwards C, Spurgeon SK, Patton RJ (2000) Sliding mode observers for fault detection and isolation. Automatica 36(4):541–553
10. Gu K, Kharitonov VL, Chen J (2003) Stability of time-delay systems, 1st edn. Birkhäuser, Boston
11. Holloway J, Krstic M (2016) Predictor observers for proportional navigation systems subjected to seeker delay. IEEE Trans Contr Syst Technol 24(6):2002–2015
12. Hui S, Zak S (2005) Observer design for systems with unknown inputs. Int J Appl Math Comput Sci 15:431–446
13. Krstic M (2010) Compensation of infinite-dimensional actuator and sensor dynamics. IEEE Contr Syst Mag 30(1):22–41
14. Krstic M (2010) Lyapunov stability of linear predictor feedback for time-varying input delay. IEEE Trans Automat Contr 55(2):554–559
15. Léchappé V, Moulay E, Plestan F, Glumineau A, Chriette A (2015) New predictive scheme for the control of LTI systems with input delay and unknown disturbances. Automatica 52:179–184
16. Léchappé V, Moulay E, Plestan F (2018) Prediction-based control of LTI systems with input and output time-varying delays. Syst Control Lett 112:24–30
17. Levant A (1993) Sliding order and sliding accuracy in sliding mode control. Int J Control 58(6):1247–1263
18. Mirkin L, Raskin N (2003) Every stabilizing dead-time controller has an observer-predictor structure. Automatica 39(10):1747–1754
19. Moreno JA, Osorio M (2008) A Lyapunov approach to second-order sliding mode controllers and observers. In: 2008 47th IEEE conference on decision and control, pp 2856–2861
20. Nagesh I, Edwards C (2014) A multivariable super-twisting sliding mode approach. Automatica 50:984–988
21. Patton RJ, Chen J (1993) Optimal unknown input distribution matrix selection in robust fault diagnosis. Automatica 29(4):837–841
22. Pinto HLCP, Oliveira TR, Hsu L, Krstic M (2018) Sliding mode control for disturbance rejection in systems with measurement delay using PDE-backstepping predictor. In: The 2018 American Control Conference, pp 4099–4104
23. Shtessel Y, Edwards C, Fridman L, Levant A (2014) Sliding mode control and observation, 1st edn. Springer, New York

24. Smith OJM (1957) Closer control of loops with dead time. Chem Eng Progr 53(5):217–219
25. Smyshlyaev A, Krstic M (2005) Backstepping observers for a class of parabolic PDEs. Syst Control Lett 54(7):613–625
26. Strang G (2009) Introduction to linear algebra, 4th edn. Wellesley Cambridge Press, Wellesley
27. Tan CP, Edwards C (2003) Sliding mode observers for robust detection and reconstruction of actuator and sensor faults. Int J Robust Nonlin Control 13(5):443–463
28. Utkin VI (1992) Sliding modes in control and optimization. Springer, Berlin

Part III
Stabilization

Chapter 5
Stabilization via Norm-State and Cascade High-Gain Observers for State-Output Delays

This final chapter, which constitutes Part III of this brief, introduces a sliding-mode control strategy to address the stabilization problem of nonlinear systems with multiple time delays in state variables and output signals, rather than focusing on fault reconstruction (Part I) or disturbance rejection (Part II).

In this novel recipe, we replace the PDE-backstepping predictor employed in Chaps. 3 and 4 by simply combining an asymptotic norm-state observer for the zero dynamics with cascaded high-gain observers, thus enabling the estimation of unmeasured states within a chain of integrators influenced by nonlinear input disturbances. One can theoretically achieve stability for any given delay value by constructing a sufficiently large number of observers, with the latter being the case of interest here. Global stabilization is achieved in the closed-loop system using only output feedback, eliminating chattering effects typically associated with sliding-mode control. Simulation results verify the method's efficacy across various scenarios, including an application to missile guidance under seeker delay conditions.

5.1 Problem Statement

Consider the class of nonlinear systems with multiple time delays

$$\dot{\eta}(t) = A_0\eta + f_0(\eta, t) + f_1(\eta(t - d_\eta(t)), t) + f_2(y, t)\,, \tag{5.1}$$

$$\dot{x}(t) = Ax(t) + B\left[k_p(t)u + w(\eta, x, t)\right]\,, \tag{5.2}$$

$$y(t) = Cx(t - d_x)\,, \tag{5.3}$$

H. L. C. P. Pinto et al., *Sliding-Mode Control and Estimation for Time-Delay Systems*, SpringerBriefs in Control, Automation and Robotics,
https://doi.org/10.1007/978-3-032-30178-9_5

where $u \in \mathbb{R}$ is the control input, $y \in \mathbb{R}$ is the measured output, and $x \in \mathbb{R}^l$ and $\eta \in \mathbb{R}^{n-l}$ are unmeasured state vectors. The matrices A_0, A, B, and C with appropriate dimensions, and the functions $f_0(\cdot,\cdot)$, $f_1(\cdot,\cdot)$, $f_2(\cdot,\cdot)$, $k_p(\cdot)$, and $w(\cdot,\cdot,\cdot)$ are known.

For convenience, the subscript d is introduced to denote the time-delayed signals [38]

$$\eta_d(t) := \eta(t - d_\eta(t)), \qquad x_d(t) := x(t - d_x). \tag{5.4}$$

Multiple time delays are allowed and considered known, in the state (d_η) and in the output signal (d_x). The initial conditions are given by

$$\eta(t) = \eta_0(t), \quad x(t) = x_0(t), \quad t \in [-d_{\max}, 0], \tag{5.5}$$

where $\eta_0(t)$ and $x_0(t)$ are vector functions continuous in $t \in [-d_{\max}, 0]$, and $d_{\max} := \max\{d_x, \sup_{t\in\mathbb{R}_+}(d_\eta(t))\}$ is the maximum allowed time delay.

The triple

$$A = \begin{bmatrix} 0 & 1 & 0 & \cdots & 0 \\ 0 & 0 & 1 & \ddots & \vdots \\ 0 & 0 & 0 & \ddots & 0 \\ \vdots & \vdots & \ddots & \ddots & 1 \\ 0 & 0 & \cdots & 0 & 0 \end{bmatrix}, \quad B = \begin{bmatrix} 0 \\ \vdots \\ 0 \\ 1 \end{bmatrix}, \quad C = \begin{bmatrix} 1 & 0 & \cdots & 0 \end{bmatrix}, \tag{5.6}$$

is in the Brunovsky's controller form, *i.e.*, it represents a chain of l integrators [20, p. 514].

The following assumptions should be satisfied:

(A1) The time delay $d_\eta(t)$ is a piecewise-continuous function bounded by $0 < \underline{d} \leq d_\eta(t) \leq \bar{d} < +\infty$, $\forall t$. The time delay $d_x > 0$ in the output signal is constant.

(A2) The matrix A_0 is Hurwitz with *stability margin* satisfying

$$\gamma_0 < -\max_i\{\mathrm{Re}(\gamma_i)\}, \tag{5.7}$$

where $\{\gamma_i\}$ are the eigenvalues of A_0, and a constant $c_1 \geq 1$ is known such that [8]

$$\|e^{A_0 t}\| \leq c_1 e^{\gamma_0 t}, \quad \forall t \geq 0. \tag{5.8}$$

(A3) $\exists \varphi_f(\cdot) \in \mathcal{K}_\infty$ such that $\|f_2(y,t)\| \leq \varphi_f(|y|)$, $\forall y \in \mathbb{R}$, $\forall t \in \mathbb{R}_+$. The output dependent nonlinear function $f_2(y,t)$ is piecewise continuous in t, locally Lipschitz, and continuous with respect to y.

(A4) The nonlinear functions $f_0(\eta, t)$ and $f_1(\eta_d, t)$ are piecewise continuous in t, $f_0(0, t) = f_1(0, t) = 0$, $\forall t \in \mathbb{R}_+$, globally Lipschitz and continuous in the other arguments, satisfying

$$\|f_0(\eta, t) - f_0(\eta', t)\| \leq \mu_0 \|\eta - \eta'\| ,$$
$$\|f_1(\eta_d, t) - f_1(\eta_d', t)\| \leq \mu_1 \|\eta_d - \eta_d'\| , \tag{5.9}$$

$\forall (\eta, \eta') \in \mathbb{R}^{n-l} \times \mathbb{R}^{n-l}, \forall (\eta_d, \eta_d') \in \mathbb{R}^{n-l} \times \mathbb{R}^{n-l}, \forall t \in \mathbb{R}_+$, where $\mu_0, \mu_1 \geq 0$ are known constants.

(A5) The function $k_p(t)$ is piecewise continuous in t, its sign is known and constant, and $|k_p(t)| \geq \underline{k}_p > 0$, $\forall t \in \mathbb{R}_+$, where $\underline{k}_p$ is a known constant lower bound for $|k_p(t)|$.

(A6) The function $w(\eta, x, t)$ is globally Lipschitz, uniformly in t with respect to η and x, *i.e.*, there exists a constant $\beta > 0$ such that the inequality

$$\left|w(\eta, x, t) - w(\eta', x', t)\right| \leq \beta \left[\left\|\eta - \eta'\right\| + \left\|x - x'\right\|\right] \tag{5.10}$$

is verified $\forall (\eta, \eta') \in \mathbb{R}^{n-l} \times \mathbb{R}^{n-l}, \forall (x, x') \in \mathbb{R}^l \times \mathbb{R}^l, \forall t \in \mathbb{R}_+$. In addition, constants $k_\eta, k_x \geq 0$ and a piecewise-continuous uniformly bounded scalar function $\varphi_w(t) \geq 0$ are known such that the upper bound

$$|w(\eta, x, t)| \leq k_\eta \|\eta(t)\| + k_x \|x(t)\| + \varphi_w(t) \tag{5.11}$$

is valid $\forall \eta \in \mathbb{R}^{n-l}, \forall x \in \mathbb{R}^l, \forall t \in \mathbb{R}_+$.

Assumption **(A1)** characterizes the multiple time delays coped with in this chapter. The time delay $d_\eta(t)$ is allowed to be time varying, and there is no limitation on the rate of change of the time delay $(\dot{d}_\eta(t))$. From this point of view, assumption **(A1)** is less restrictive than the condition $\dot{d}_\eta(t) < 1$ assumed in [18, Theorem 5.2]. On the other hand, the output time delay d_x should be constant, but it can be arbitrarily long.

State observers can be applied in output-feedback controllers, e.g., [26, 38]. However, their design seems to be difficult for the class of time-varying nonlinear systems considered here, and they do not take into account time-delayed output. In contrast, Assumptions **(A1)**, **(A2)**, **(A3)**, and **(A4)** provide conditions needed for the development of an observer for the unmeasured state vector η. Note that the Hurwitz assumption **(A2)** on A_0 can be relaxed following the concepts discussed by [6, Sec. 3.3].

Assumption **(A5)** and inequality (5.11) in **(A6)** are used for the synthesis of the modulation signal in the output-feedback control law. Particularly, the globally Lipschitz condition on the nonlinear function $w(\eta, x, t)$ assumed in **(A6)** is needed for the cascade observer design applied to estimate the state vector x. Inequality (5.10) with $x' \equiv 0$ and $\eta' \equiv 0$ also implies the upper bound given in (5.11), since $\varphi_w(t) \geq |w(0, 0, t)|$. Note that the nonlinear terms in (5.1)–(5.2) are not uniformly bounded. In practice, systems with unbounded physical variables can be coped with the proposed control scheme, such as velocity or currents in motor drives. Indeed, the globally Lipschitz assumption is a boundedness condition for derivatives, such as acceleration and current rate, which are bounded by natural restrictions in industrial applications. Moreover, this class of plants includes many applications, for example: bioreactors, networks, and teleoperation.

The control objective is to guarantee global stability properties of the closed-loop system. An output-feedback dynamic control law u will be designed such that the state vectors η and x converge at least asymptotically to zero, and all signals in the closed-loop system remain uniformly bounded.

5.2 Observer for the Zero Dynamics State

To design the control law, here is proposed the following observer for the state η of the subsystem (5.1):

$$\begin{aligned}\dot{\hat{\eta}}(t) = A_0\hat{\eta}(t) + f_0(\hat{\eta}(t), t) + \\ + f_1(\hat{\eta}(t - d_\eta(t)), t) + f_2(y, t)\,,\end{aligned} \tag{5.12}$$

where $\hat{\eta} \in \mathbb{R}^{n-l}$ is the estimated state vector. To analyze this observer, the estimation error

$$\tilde{\eta} := \eta - \hat{\eta} \tag{5.13}$$

is introduced. Therefore, subtracting equation (5.12) from (5.1), the dynamic equation of the estimation error can be obtained:

$$\begin{aligned}\dot{\tilde{\eta}}(t) = A_0\tilde{\eta}(t) + f_0(\eta(t), t) - f_0(\hat{\eta}(t), t) + \\ + f_1(\eta(t - d_\eta(t)), t) - f_1(\hat{\eta}(t - d_\eta(t)), t)\,,\end{aligned} \tag{5.14}$$

with initial condition

$$\tilde{\eta}(t) = \tilde{\eta}_0(t)\,, \quad t \in [-d_{\max}, 0]\,. \tag{5.15}$$

The solution of this differential equation is given by $t \geq 0$:

$$\begin{aligned}\tilde{\eta}(t) = \mathrm{e}^{A_0 t}\tilde{\eta}_0(0) + \mathrm{e}^{A_0 t} * [f_0(\eta(t), t) - f_0(\hat{\eta}(t), t) \\ + f_1(\eta(t - d_\eta(t)), t) - f_1(\hat{\eta}(t - d_\eta(t)), t)]\,.\end{aligned} \tag{5.16}$$

From assumptions **(A2)** and **(A4)**, the norm of the estimation error can be bounded by

$$\begin{aligned}\|\tilde{\eta}(t)\| &\leq c_1\tilde{\eta}_0^* \mathrm{e}^{-\gamma_0 t} + c_1 \mathrm{e}^{-\gamma_0 t} * \\ &\quad * [\| f_0(\eta(t), t) - f_0(\hat{\eta}(t), t)\| + \\ &\quad + \| f_1(\eta(t - d_\eta(t)), t) - f_1(\hat{\eta}(t - d_\eta(t)), t)\|] \\ &\leq c_1\tilde{\eta}_0^* \mathrm{e}^{-\gamma_0 t} + c_1 \mathrm{e}^{-\gamma_0 t} * \left[\mu_0 \|\tilde{\eta}(t)\| + \right. \\ &\quad \left. + \mu_1 \|\tilde{\eta}(t - d_\eta(t))\|\right] \\ &\leq r(t)\,,\end{aligned} \tag{5.17}$$

where

$$r(t) := c_1 \mathrm{e}^{-\gamma_0 t} * \left[\mu_0 \|\tilde{\eta}\| + \mu_1 \tilde{\eta}_{\mathrm{sup}}\right] + c_1 \tilde{\eta}_0^* \mathrm{e}^{-\gamma_0 t}\,, \tag{5.18}$$

$$\tilde{\eta}_0^* := \sup_{t \in [-d_{\max}, 0]} \|\tilde{\eta}_0(t)\|\,, \tag{5.19}$$

$$\tilde{\eta}_{\mathrm{sup}}(t) := \sup_{\tau \in [\underline{d}, \bar{d}]} \|\tilde{\eta}(t - \tau)\|\,. \tag{5.20}$$

The constants $c_1 \geq 1$ and $\gamma_0 > 0$ must satisfy assumption **(A2)**. For design purposes, these constants can be computed for the matrix Λ_0 using the method proposed by [8].

The function $r(t)$ given by (5.18) is the solution of the differential equation

$$\begin{aligned}\dot{r}(t) &= -\gamma_0 r(t) + c_1 \left[\mu_0 \|\tilde{\eta}(t)\| + \mu_1 \tilde{\eta}_{\mathrm{sup}}(t)\right]\,, \quad t \geq 0\,, \\ r(t) &= c_1 \tilde{\eta}_0^*, \quad \forall t \in [-d_{\max}, 0]\,.\end{aligned} \tag{5.21}$$

Since $r(t) \geq \|\tilde{\eta}(t)\|$, $\forall t \geq -\bar{d}$, then

$$\tilde{\eta}_{\mathrm{sup}}(t) \leq \sup_{\tau \in [\underline{d}, \bar{d}]} r(t - \tau)\,, \quad \forall t \geq 0\,. \tag{5.22}$$

In addition, $\dot{r} \geq -\gamma_0 r$, and by using the comparison theorem [13, Theorem 7], one has from (5.22) that, $\forall t \geq 0$,

$$\tilde{\eta}_{\mathrm{sup}}(t) \leq \sup_{\tau \in [\underline{d}, \bar{d}]} \mathrm{e}^{-\gamma_0 [(t - \tau) - t]} r(t) \leq \mathrm{e}^{\gamma_0 \bar{d}} r(t)\,. \tag{5.23}$$

Upon substituting $\|\tilde{\eta}\|$ by r and $\tilde{\eta}_{\text{sup}}$ by $e^{\gamma_0 \bar{d}} r$ in (5.21), the differential equation

$$\dot{\bar{r}} = (c_1\mu_0 + c_1\mu_1 e^{\gamma_0 \bar{d}} - \gamma_0)\bar{r} \,,$$
$$\bar{r}(0) = r(0) = c_1\tilde{\eta}_0^* \,, \tag{5.24}$$

can be obtained. The solution of this equation satisfies $\bar{r}(t) \geq r(t) \geq \|\tilde{\eta}(t)\|$, $\forall t \geq 0$. So, it can be concluded that, if

$$\lambda_0 := \gamma_0 - c_1\mu_0 - c_1\mu_1 e^{\gamma_0 \bar{d}} > 0 \,, \tag{5.25}$$

then (5.24) and, consequently, (5.14) are exponentially stable, i.e.,

$$\|\tilde{\eta}(t)\| \leq c_1\tilde{\eta}_0^* e^{-\lambda_0 t} \,, \quad t \geq 0 \,, \tag{5.26}$$

and $\tilde{\eta}(t) \to 0$ as $t \to +\infty$.

Note that the condition (5.25) on λ_0 implies the transcendental inequality

$$\gamma_0 > c_1\mu_0 + c_1\mu_1 e^{\gamma_0 \bar{d}} \,. \tag{5.27}$$

Such inequality is useful to determine the maximum allowed delay. Rewriting (5.27) as $e^{\gamma_0 \bar{d}} \leq (\gamma_0 - c_1\mu_0)/(c_1\mu_1)$ and then applying $\ln(\cdot)$ in both sides yield

$$\bar{d} \leq \frac{1}{\gamma_0} \ln \left(\frac{\gamma_0 - c_1\mu_0}{c_1\mu_1} \right) . \tag{5.28}$$

Remark 5.1 As in [25], we can note that the maximum time delay is limited by the stability margin γ_0 of A_0, and there is a natural limitation imposed by the argument of $\ln(\cdot)$ function in (5.28), which must be positive.

Remark 5.2 Assumption **(A3)** and the condition (5.25) imply that the zero dynamics of (5.1) with $y \equiv 0$ is exponentially stable. Moreover, this subsystem is OSS, as can be concluded from the analysis presented by Coutinho et al. [6].

5.3 Cascade High-Gain Observers

In this section, the cascaded high-gain observers proposed by Ahmed-Ali et al. [1] are briefly reviewed. These observers will be used to estimate the state x of the nonlinear subsystem (5.2) with delayed output signal (5.3).

According to [1], the nonlinear function

$$\phi(\eta, x, u, t) := B\left[k_p(t)u + w(\eta, x, t)\right] \tag{5.29}$$

should be globally Lipschitz with respect to η and x and uniformly in u and t, which is ensured by assumptions **(A5)** and **(A6)**.

The analysis presented in [1] had not considered disturbances in the nonlinear function $\phi(\eta, x, u, t)$. However, it is possible to show that such cascade observers are robust to exponentially decaying terms bounded by (5.26), which result from the state $\hat{\eta}$ estimated by the observer (5.12). Therefore, it can be concluded from the Lipschitz assumption **(A6)** that, $\forall t \geq 0$,

$$\left|\phi(\eta, x, u, t) - \phi(\hat{\eta}, x, u, t)\right| \leq \beta \left\|\eta - \hat{\eta}\right\| \leq \beta\, c_1 \tilde{\eta}_0^* \mathrm{e}^{-\lambda_0 t}. \tag{5.30}$$

Thus, here the nonlinear function $\phi(\eta, x, u, t)$ can be estimated by $\phi(\hat{\eta}, x, u, t)$ for the cascade observer and control law synthesis.

5.3.1 Analysis of a High-Gain Observer for Systems with Small Time Delay

If the time delay d_x is sufficiently small, only one high-gain observer would be needed to estimate the state x. The exponential convergence of the estimated state is presented in the following lemma in [1, Lemma 1] using a Lyapunov–Krasovskii functional [14].

Lemma 5.1 *Consider the following observer:*

$$\dot{\hat{x}} = A\hat{x} + \phi(\hat{\eta}, \hat{x}, u, t) - \theta \Delta^{-1} K (C\hat{x}(t - d_x) - y), \tag{5.31}$$

where the vector $K \in \mathbb{R}^l$ is chosen such that the matrix $A - KC$ is Hurwitz and

$$\Delta = \operatorname{diag}\left\{1, \ldots, \frac{1}{\theta^{i-1}}, \ldots, \frac{1}{\theta^{l-1}}\right\}. \tag{5.32}$$

Then, for sufficiently large θ, $\exists d_1 > 0$ such that if $d_x \in [0, d_1]$, the observer state $\hat{x}$ (5.31) converges exponentially to the state x of the subsystem (5.2)–(5.3).

In [1, Lemma 1], the maximum delay admitted in the observer (5.31)

$$d_1 = \frac{\mu}{\theta^2} \tag{5.33}$$

ensures $\hat{x}(t) \rightarrow x(t)$, provided the parameter $\theta \geq 1$ is chosen to verify the inequalities

$$\begin{aligned} \frac{\mu\theta}{2} &> 2\lambda_{\max}(P)\sqrt{l}\beta + 2\mu\left[\|A - KC\| + 2\lambda_{\max}(P)\sqrt{l}\beta\right]^2, \\ \theta^2 &> 2\lambda_{\max}^2(P)\,\|K\|^2\,\theta + 2\mu^2\,\|K\|^2, \end{aligned} \tag{5.34}$$

where $P = P^T > 0$ satisfies the algebraic Lyapunov inequality

$$P(A - KC) + (A - KC)^T P \leq -\mu I \,, \tag{5.35}$$

and $\mu > 0$ is a design parameter.

In the next section, observers connected in cascade will be introduced to deal with larger delays.

5.3.2 Design of High-Gain Observers Connected in Cascade

The cascade connection of observers is an approach to deal with larger delays [1, 15] using the idea of delay distribution [24, Remark 7.8]. The number of observers (m) is proportional to the time delay d_x and should satisfy [1]:

$$m \geq \frac{d_x}{d_1} \,, \tag{5.36}$$

where $d_1 > 0$ is the maximum delay admitted by a single-stage high-gain observer.

Each observer ($j = 1, \ldots, m$) estimates a delayed state vector

$$\hat{x}_j(t) = \hat{x}\left(t - d_x + j\frac{d_x}{m}\right) \,, \tag{5.37}$$

where the delay is equally distributed among all observers (d_x/m). The following notation represents the delayed control signal u and the delayed state $\hat{\eta}$ estimated by the observer (5.12):

$$u_j(t) = u\left(t - d_x + j\frac{d_x}{m}\right) \,, \quad \hat{\eta}_j(t) = \hat{\eta}\left(t - d_x + j\frac{d_x}{m}\right) \,. \tag{5.38}$$

Thus, the observers connected in cascade are given by

$$\begin{aligned}
\dot{\hat{x}}_1 &= A\hat{x}_1 + \phi\left(\hat{\eta}_1, \hat{x}_1, u_1, t - d_x + \frac{d_x}{m}\right) \\
&\quad - \theta\Delta^{-1}K\left[C\hat{x}_1\left(t - \frac{d_x}{m}\right) - y\right] \,, \\
\hat{y}_1 &= C\hat{x}_1 \,, \\
&\vdots \\
\dot{\hat{x}}_j &= A\hat{x}_j + \phi\left(\hat{\eta}_j, \hat{x}_j, u_j, t - d_x + j\frac{d_x}{m}\right)
\end{aligned}$$

$$
\begin{aligned}
&\quad - \theta\Delta^{-1}K\left[C\hat{x}_j\left(t-\frac{d_x}{m}\right)-\hat{y}_{j-1}\right], \\
\hat{y}_j &= C\hat{x}_j\,, \\
&\vdots \\
\dot{\hat{x}}_m &= A\hat{x}_m + \phi\left(\hat{\eta}_m, \hat{x}_m, u, t\right) \\
&\quad - \theta\Delta^{-1}K\left[C\hat{x}_m\left(t-\frac{d_x}{m}\right)-\hat{y}_{m-1}\right], \\
\hat{y}_m &= C\hat{x}_m\,,
\end{aligned}
\tag{5.39}
$$

where the vector $K \in \mathbb{R}^l$ is chosen such that the matrix $A - KC$ is Hurwitz, $\theta > 0$ is the high gain, and Δ defined in (5.32).

Remark 5.3 The vector $\hat{x}_j$ is an estimate of the delayed state $x_j(t) := x\left(t-d_x+j\frac{d_x}{m}\right)$. The state vector $\hat{x}(t) := \hat{x}_m(t)$ is an estimate of the current state $x(t)$ of the time-delay subsystem (5.2)–(5.3).

5.3.3 *Analysis of High-Gain Observers Connected in Cascade*

Ahmed-Ali et al. [1] had shown that a single-stage observer is sufficient to provide convergence of the estimated state for systems with a small time delay d_1 in the output signal. Furthermore, it had also been proved that a sufficient number of cascade high-gain observers can estimate the state of the system (5.2)–(5.3), for arbitrarily long and constant time delay. This is stated in [1, Theorem 1], rewritten below.

Lemma 5.2 *Consider the system described in (5.2)–(5.3). Then, for any constant and known delay d_x, there exist a sufficiently large positive constant θ and an integer m such that the state estimated by the last observer $\hat{x}_m$ in (5.39) converges exponentially toward the state x of the system (5.2)–(5.3). Moreover, the state estimation errors $\tilde{x}_j(t) := x_j(t) - \hat{x}_j(t)$, $\forall j \in \{1, \ldots, m\}$, converge exponentially to zero.*

5.4 Output-Feedback Sliding-Mode Control

Here, we propose the sliding-mode controller that applies the observer (5.12) for the state delayed subsystem (5.1) and high-gain cascade observers (5.39) for the output delayed subsystem (5.2)–(5.3) discussed above.

To design the output-feedback SMC, consider the system (5.2)–(5.3). Then, substituting u by $u + K_m x/k_p(t) - K_m x/k_p(t)$, one has

$$\dot{x} = A_m x + B\left[k_p(t)u + w(\eta, x, t) - K_m x\right], \tag{5.40}$$

$$y = Cx(t - d_x), \tag{5.41}$$

where $x(t - d_x) = [y, \dot{y}, \ldots, y^{(l-1)}]^T$ and $A_m := A + BK_m$. The matrix $K_m := -[a_0, \ldots, a_{l-1}]$ is chosen such that the characteristic polynomial

$$p(s) = s^l + a_{l-1}s^{l-1} + \ldots + a_0 \tag{5.42}$$

of the matrix A_m is Hurwitz.

The state equation (5.40) can be rewritten as

$$\dot{x} = A_m x + Bk_p(t)\left[u + w_u(\eta, x, t)\right], \tag{5.43}$$

where the equivalent input disturbance is given by

$$w_u(\eta, x, t) = \left[w(\eta, x, t) - K_m x\right]/k_p(t). \tag{5.44}$$

If the full-state vector x were available for feedback, one could choose $\sigma = Sx = 0$ as the ideal sliding surface. On the other hand, since only the system output signal y is available for feedback, the sliding surface can be chosen as [9]

$$\hat{\sigma} := S\hat{x} = 0, \qquad S := \left[b_0 \cdots b_{l-2}\ 1\right], \tag{5.45}$$

where $b_0, \ldots, b_{l-2}, 1$ are coefficients of the Hurwitz polynomial

$$p_S(s) = s^{l-1} + b_{l-2}s^{l-2} + \ldots + b_0, \tag{5.46}$$

such that the transfer function $p_S(s)/p(s) = S(sI - A_m)^{-1}B$ is strictly positive real (SPR) [20, Section 6.3]. The signal $\hat{x} := \hat{x}_m$ is the estimate of the system state x via observer (5.39).

The proposed control law is

$$u = -\text{sgn}(k_p)\varrho(\hat{\eta}, \hat{x}, t)\text{sgn}(\hat{\sigma}(t)), \tag{5.47}$$

where the modulation function $\varrho(\hat{\eta}, \hat{x}, t)$ is a nonnegative scalar function absolutely continuous in $\hat{\eta}$ and $\hat{x}$ and piecewise continuous and bounded in t for each fixed $\hat{\eta}, \hat{x}$. It will be shown that if the inequality

$$\varrho(\hat{\eta}, \hat{x}, t) \geq |w_u(\eta, x, t)| + \delta \tag{5.48}$$

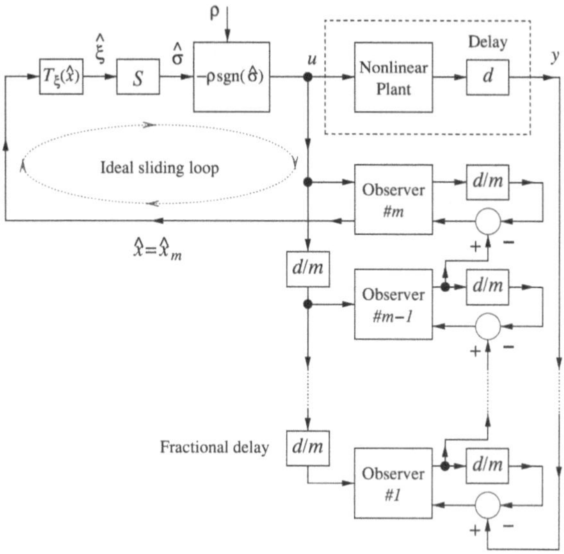

Fig. 5.1 Block diagram of sliding-mode controller based on high-gain observers connected in cascade for nonlinear time-delay systems. The synthesis of the modulation signal $\varrho(\hat{\eta}, \hat{x}, t)$ is not shown to avoid clutter. Here, the output delay is redefined as $d := d_x$ to simplify the notation in the block diagram illustration

is satisfied *modulo* exponentially decaying terms, then global stabilization can be guaranteed. The parameter $\delta \geq 0$ is an arbitrary constant. For instance, recalling assumptions **(A5)** and **(A6)** (eq. (5.11)), a function that satisfies (5.48) is

$$\varrho(\hat{\eta}, \hat{x}, t) = \left[\left| K_m \hat{x} \right| + k_\eta \|\hat{\eta}\| + k_x \|\hat{x}\| + \varphi_w(t) \right] / \underline{k}_p + \delta \,. \tag{5.49}$$

The sliding-mode controller based on HGOs connected in cascade for nonlinear time-delay systems is described by the block diagram in Fig. 5.1.

5.5 Closed-Loop Stability Analysis

The following lemma shows that the system (5.43)–(5.44) with state vector x and output signal $\tilde{\sigma} := S\tilde{x}$ is OSS, where the state estimation error is defined as

$$\tilde{x} := x - \hat{x} \,. \tag{5.50}$$

Therefore, according to the next lemma, it will be demonstrated that if $\|\tilde{x}(t)\| \to 0$, then $\|x(t)\| \to 0$.

Lemma 5.3 (OSS Property from $S\tilde{x}$ to x) *Consider the dynamic system (5.43)–(5.44) with state vector x, output signal $\tilde{\sigma} = S\tilde{x} = Sx - S\hat{x}$, and the control law u (5.47) with modulation function ϱ (5.49). Then, (5.43)–(5.44) is OSS with respect* *to $\tilde{\sigma}$ and $\exists k_e > 0$ such that the following inequality holds*

$$\|x(t)\| \leq k_e|S\tilde{x}(t)| + \pi_e(t)\,, \quad \forall t \geq 0\,, \tag{5.51}$$

where $\pi_e(t)$ is an exponentially decaying term dependent on the initial conditions.

Proof In the following, k_i's denote appropriate positive constants and π_i's denote exponentially decaying terms that depend on the initial conditions of the closed-loop system.

Introduce the transformation

$$\bar{x} = Tx\,, \quad T = \begin{bmatrix} 1 & 0 & 0 & \cdots & 0 \\ 0 & 1 & 0 & \cdots & 0 \\ 0 & 0 & 1 & \ddots & \vdots \\ \vdots & \vdots & \cdots & \ddots & 0 \\ b_0 & b_1 & \cdots & b_{l-2} & 1 \end{bmatrix}, \tag{5.52}$$

and the auxiliary signal $\sigma = Sx$ as in (5.45), such that the system (5.43) can be represented in the normal form [20, Section 13.2] with state

$$\bar{x}(t-d_x) = \left[y\ \dot{y} \cdots y^{(l-2)}\ \sigma(x(t-d_x))\right]^T.$$

Since the triple $\{A_m, B, S\}$ (see (5.45) and (5.46)) is SPR, one can conclude that (5.43) is OSS from the output σ to the state $\bar{x}$. Therefore, $\bar{x}$ satisfies $\|\bar{x}\| \leq k_1|\sigma| + \pi_1(t)$. Since from (5.52), $\|x\| \leq \|T^{-1}\|\,\|\bar{x}\|$, one has

$$\|x\| \leq k_2|Sx| + \pi_2(t)\,. \tag{5.53}$$

In what follows, two cases will be considered:

(i) In the first case, $|Sx| > |S\tilde{x}|$ implies $\text{sgn}(\hat{\sigma}) = \text{sgn}(Sx)$. From the Kalman–Yakubovich lemma [20, Sec. 6.3], there exist $P = P^T > 0$ and $Q = Q^T > 0$ matrices that satisfy $A_m^T P + PA_m = -Q$ and $B^T P = S$. Using a Lyapunov function candidate $V = x^T Px$, one concludes that the time derivative of V along the solutions in (5.43) satisfies $\dot{V} \leq -\lambda_{\min}(Q)\|x\|^2 - 2k_p(t)|B^T Px|[\varrho - |w_u|]$, or equivalently,

$$\dot{V} \leq -\lambda_{\min}(Q)\|x\|^2 - 2\,|Sx|\left[\underline{k}_p\varrho - \left|k_p(t)w_u\right|\right], \tag{5.54}$$

where $\underline{k}_p$ is the lower bound for $|k_p(t)|$ assumed in **(A5)**. Considering that the modulation function ϱ in (5.49) verifies the inequality (5.48), then $\dot{V} \leq -\lambda_{\min}(Q)\|x\|^2$. Then, one can conclude that $x \to 0$ and $Sx \to 0$ exponentially, for any initial condition.

(ii) In the second case, $|Sx| \leq |S\tilde{x}|$. This, in conjunction with case (i), leads to the conclusion that $|Sx| \leq |S\tilde{x}| + \pi_3$. Then applying the last inequality in (5.53), one concludes that $\|x\| \leq k_2|S\tilde{x}| + \pi_4(t)$. Thus the dynamics that governs x is OSS with respect to $S\tilde{x}$, according to (5.51).

□

At this point, we can state the following result for global stability, *i.e.*, the stability properties are independent of the initial conditions that are not required to be bounded. To that end, the state of the closed-loop system (including plant and observers) is defined as

$$z(t) := \left[\eta^T(t)\, x^T(t)\, \hat{\eta}^T(t)\, \hat{x}_1^T(t) \cdots \hat{x}_m^T(t)\right]^T . \tag{5.55}$$

Theorem 5.1 (Global Asymptotic Stability) *Consider the nonlinear system with multiple time delays given in (5.1)–(5.3), the control law proposed in (5.47) with modulation function ϱ given by (5.49), and the observers (5.12) and (5.39). If assumptions **(A1)**–**(A6)** and inequality (5.27) hold, then the equilibrium point $z = 0$ of the closed-loop system is globally asymptotically stable and all signals are uniformly bounded.*

Proof From Lemma 5.2, the estimated states $\hat{x}_j(t)$ converge exponentially to the states $x_j(t)$ and, consequently, the estimation errors $\tilde{x}_j(t)$, $\forall j \in \{1, \ldots, m\}$, converge exponentially to zero, *i.e.*,

$$\|\tilde{x}(t)\| \to 0\,, \quad \text{as} \quad t \to +\infty\,. \tag{5.56}$$

According to inequality (5.51) in Lemma 5.3, the norm of the state x is bounded by the estimation error norm plus an exponentially decaying term $\pi_e(t)$.

$$\|x(t)\| \leq k_e\|\tilde{x}(t)\| + \pi_e(t)\,, \tag{5.57}$$

with $k_e > 0$ being an appropriate constant. Therefore, using (5.56) in (5.51) and reminding that $\hat{x} = x - \tilde{x}$, one can conclude that $x(t)$ and $\hat{x}(t)$ tend exponentially to zero. This implies that all the delayed estimated states $\hat{x}_j(t)$ and the output signal y tend exponentially to zero. From inequality (5.26), $\tilde{\eta}(t)$ tends exponentially to zero. From Remark 5.2, $\eta(t)$ and $\hat{\eta}(t)$ tend to zero asymptotically. Consequently, $\exists \beta_z \in \mathcal{KL}$ such that the state can be bounded by

$$\|z(t)\| \leq \beta_z(z_0^*, t)\,, \quad \forall t \geq 0\,, \tag{5.58}$$

$$z_0^* := \sup_{t\in[-d_{\max},0]} \|z_0(t)\|\,, \tag{5.59}$$

where the initial condition is $z(t) = z_0(t)$, $t \in [-d_{\max}, 0]$. Since the estimated states $\hat{\eta}$ and $\hat{x}$ tend to zero and the function $\varphi_w(t)$ is uniformly bounded in view of assumption **(A6)**, the modulation function given by (5.49) is uniformly bounded.

Consequently, the control signal u and all the closed-loop system signals are also bounded. □

Corollary 5.1 (Global Exponential Stability) *In addition to the assumptions in Theorem 5.1, if* $\exists \mu_2 \geq 0$ *such that* $\| f_2(y,t)\| \leq \mu_2 \|y\|$, $\forall y, t$, *then the equilibrium point* $z = 0$ *of the closed-loop system is globally exponentially stable.*

Proof The proof follows the same steps in Theorem 5.1. Since y decays exponentially to zero, the signal $f_2(y,t)$ is also exponentially decaying. Therefore, from the OSS property of the η-dynamics, the state vectors $\eta(t)$ and $\hat{\eta}(t)$ tend to zero exponentially, completing the proof. □

Corollary 5.2 (Ideal Sliding Mode) *In addition to the assumptions in Theorem 5.1, if* $\delta > 0$ *in (5.49), then the sliding surface* $\hat{\sigma}(t) \equiv 0$ *is reached in finite time.*

Proof The dynamic equation of the last observer in (5.39) can be rewritten as

$$\begin{aligned} \dot{\hat{x}} &= Bk_p(t)\left[u + w_\phi(\hat{x},t)\right] + \\ &\quad \underbrace{+ A\hat{x} - \theta\Delta^{-1}K\left[C\hat{x}\left(t - \frac{d}{m}\right) - C\hat{x}_{m-1}\right]}_{\zeta}, \end{aligned} \tag{5.60}$$

or equivalently,

$$\dot{\hat{x}} = Bk_p(t)\left[u + w(\hat{\eta}, \hat{x}, t)\right] + \zeta\,, \tag{5.61}$$

with

$$\zeta = A\hat{x} - \theta\Delta^{-1}K\left[C\hat{x}\left(t - \frac{d}{m}\right) - C\hat{x}_{m-1}\right]. \tag{5.62}$$

Now, consider the quadratic function $V_\sigma = \hat{\sigma}^2/2$, in which $P = P^T > 0$ is the solution of

$$A_m^T P + P A_m = -Q\,, \tag{5.63}$$

where $Q = Q^T > 0$ and $PB = S^T$, reminding that the system $\{A_m, B, S\}$ is SPR.
Then, calculating $\dot{V}_\sigma$ along the solution $\hat{x}(t)$ of (5.61),

$$\dot{V}_\sigma = \hat{\sigma}\dot{\hat{\sigma}} = \hat{\sigma}S\dot{\hat{x}} = \hat{\sigma}S\left[Bk_p\left(u + w(\hat{\eta}, \hat{x}, t)\right) + \zeta\right]. \tag{5.64}$$

Since $SB = 1$ and the control signal is given by (5.47), the function $\dot{V}_\sigma$ can be rewritten as

$$\dot{V}_\sigma = \hat{\sigma}\left[k_p\left(-\text{sgn}(k_p)\varrho\,\text{sgn}(\hat{\sigma}) + w(\hat{\eta}, \hat{x}, t)\right) + S\zeta\right]$$

$$\leq |\hat{\sigma}|\, |k_p| \left[-\varrho + |w(\hat{\eta}, \hat{x}, t)| + |k_p^{-1} S \zeta| \right] . \tag{5.65}$$

Since the modulation function ϱ satisfies (5.48), the following inequality is valid

$$\dot{V}_\sigma \leq |\hat{\sigma}|\, |k_p| \left[-\delta + |k_p^{-1} S \zeta| \right] . \tag{5.66}$$

Note that, according to Theorem 5.1, x and $\hat{x}$ tend to zero. Then, there exists a finite time $T_1 > 0$ such that

$$|k_p^{-1} S \zeta| \leq \delta_1 , \quad \forall t \geq T_1 , \tag{5.67}$$

with some constant $0 < \delta_1 < \delta$. Therefore, we have that $\dot{V}_\sigma \leq |\hat{\sigma}|\, |k_p| [-\delta + \delta_1] < 0$, and the condition $\hat{\sigma}\dot{\hat{\sigma}} < 0$ for the existence of a sliding mode in some finite time is verified [36, Sec. 2.5]. □

Remark 5.4 The asymptotic observer (5.39) can eliminate chattering, even in the presence of the time delay, since the ideal sliding mode occurs in an auxiliary observer loop rather than in the main control loop, as described by [36, Section 8.3] and [9].

5.6 Numerical Results

Consider the following nonlinear system with state and output delays. To avoid clutter, here x^1 and x^2 denote the elements of the state vector x (see Remark 5.3) following the notation in [1], i.e., they are not powers of x:

$$\dot{\eta} = \begin{bmatrix} -2 & 0 \\ 0 & -1 \end{bmatrix} \eta + \begin{bmatrix} -0.5\eta_1 \sin \eta_2 + 0.2\eta_{d1} \tanh(\eta_{d2}) + y^2 \\ 0.5\eta_2 \cos \eta_1 + 0.2(\eta_{d2} - \eta_{d1}) - y^3 \end{bmatrix} , \tag{5.68}$$

$$\dot{x} = \begin{bmatrix} 0 & 1 \\ 0 & 0 \end{bmatrix} x + \begin{bmatrix} 0 \\ 1 \end{bmatrix} \left[k_p(t) u(t) + w(\eta, x, t) \right] , \tag{5.69}$$

$$y = \begin{bmatrix} 1 & 0 \end{bmatrix} x(t - d_x) , \tag{5.70}$$

where $k_p(t) \equiv 1$, $w(\eta, x, t) = 0.5x^1 + 0.5 \tanh(x^2) + 2\|\eta\|$, the state vector is $x = [x^1\ x^2]^T \in \mathbb{R}^2$, $\eta = [\eta_1\ \eta_2]^T \in \mathbb{R}^2$, and η_d is defined in (5.4).

To illustrate the need of the observer, we consider initially a scenario with constant delays where the observer is not adopted, assuming the measurement of the delayed state $x(t - d_x)$. Therefore, the state-feedback control law is given by $u = -\text{sgn}(k_p)\varrho(\eta(t), x(t - d_x), t)\text{sgn}(Sx(t - d_x))$. With $d_x = d_\eta = 0.2\,\text{s}$, it is possible to note chattering in the control signal (see Fig. 5.2b) due to the finite switching frequency imposed by the delayed measurement. As a consequence, the

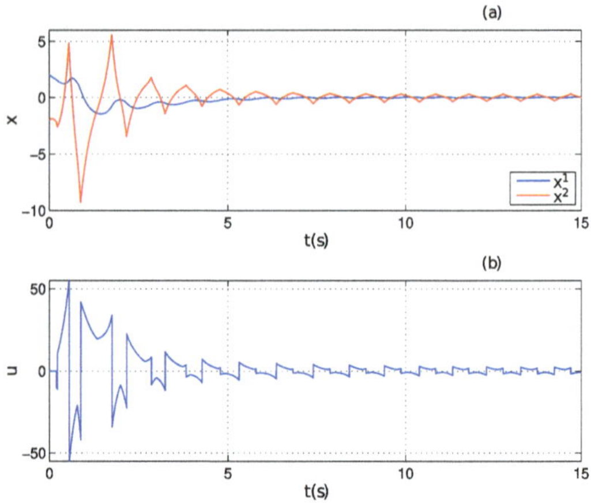

Fig. 5.2 Unstable state-feedback control with $d_x = d_\eta = 0.4$ s: actual states x^1 and x^2 (**a**) and control signal (**b**)

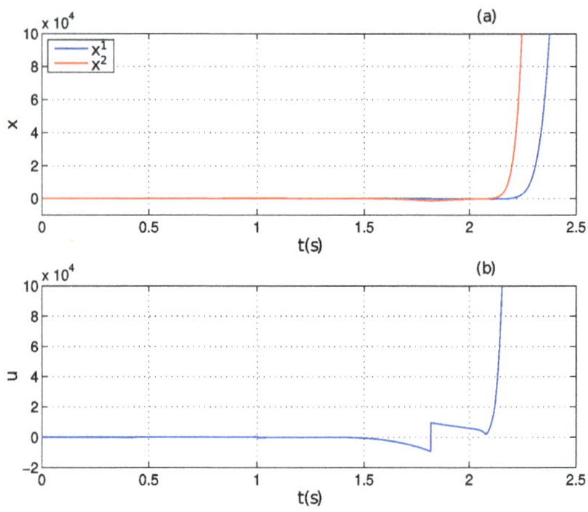

Fig. 5.3 State-feedback control with $d_x = d_\eta = 0.2$ s: actual states x^1 and x^2 (**a**) and control signal (**b**)

states are not confined to the sliding surface and asymptotic convergence is not observed in Fig. 5.2a.

For larger time delays, $d_x = d_\eta = 0.4$ s, the system becomes unstable, as can be seen in Fig. 5.3.

To illustrate the role of the observer, the parameters of the cascade observers (5.39) are set to $\theta = 4$ and $K = [0.85\ 0.24]^T$. Since the considered output delay is $d_x = 0.4$ s, then one observer would not be sufficient to estimate the state of subsystem (5.69)–(5.70). In this case, the number of observers connected in cascade is $m = 2$.

The nonlinear terms in η and in η_d presented in (5.68) satisfy (5.9) with $\mu_0 = 0.5$ and $\mu_1 = 0.2$.

On the other hand, the time-varying state delay $d_\eta(t) = 0.3\sin(5t) + 0.5$ (in seconds) is used in the control system. Its upper bound $\bar{d} = 0.8$ s along with $\gamma_0 = 1$ rad/s and $c_1 = 1$ verifies inequality (5.28), required to guarantee the stability of the observer (5.12) for the zero dynamics state.

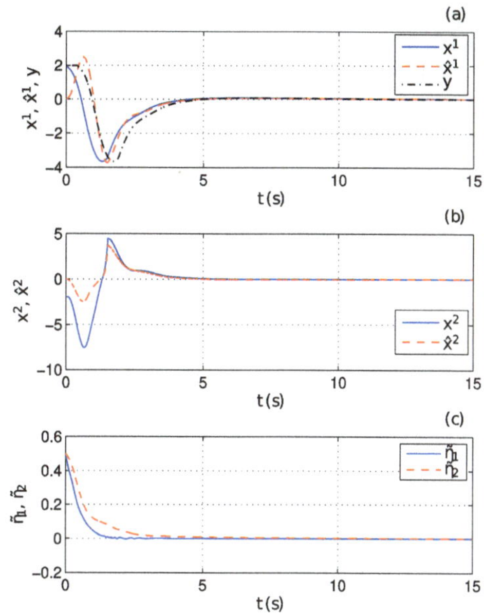

Fig. 5.4 Actual states x^1 (**a**) and x^2 (**b**), estimated states $\hat{x}^1$ (**a**) and $\hat{x}^2$ (**b**), delayed output y (**a**), and estimation errors $\tilde{\eta}_1$ and $\tilde{\eta}_2$ for the zero dynamics state (**c**)

To design the control law (5.47), the matrix $S = [11]$ is chosen to define the sliding surface as in (5.45). The modulation function ϱ is given by (5.49), where $K_m = [-2 \ -3]$. The function $\varphi_w(t) = 0.6$ and the constants $k_x = 0.6$, $k_\eta = 2.1$ were chosen to satisfy (5.11) in **(A6)**, i.e., $|w(\eta, x, t)| \leq 2.1 \|\eta\| + 0.6 \|x\| + 0.6$. The constant $\underline{k}_p = 0.9$ was chosen to satisfy **(A5)**, since $k_p(t) \geq \underline{k}_p$. The parameter $\delta = 0.1$ was arbitrarily set as in Corollary 5.2 to guarantee the existence of the sliding mode in finite time.

Figures 5.4, 5.5, and 5.6 display simulation results for this control system. The following initial conditions were considered

$$x(t) = \begin{bmatrix} 2 \\ -2 \end{bmatrix}, \ \hat{x}_j(t) = \begin{bmatrix} 0 \\ 0 \end{bmatrix}, \ \eta(t) = \begin{bmatrix} 0.5 \\ 0.5 \end{bmatrix}, \ \hat{\eta}(t) = \begin{bmatrix} 0 \\ 0 \end{bmatrix}, \tag{5.71}$$

$\forall t \in [-0.8, 0]$ and $j = 1, 2$, for the system and all observers in the simulations. The state variables η_1, η_2, x^1, and x^2 reach the equilibrium point in the origin, as expected (see Fig. 5.4). Moreover, the estimation errors $\tilde{x}$ and $\tilde{\eta}$ converge to zero, as can also be concluded from Fig. 5.4.

To illustrate the effect of time-varying delay, the actual states (η_1, η_2) of the subsystem (5.68) and the corresponding delayed state variables (η_{d1}, η_{d2}) are shown in Fig. 5.5. The delayed signals are distorted due to the time-varying nature of the delay $d_\eta(t)$.

In Fig. 5.6a, it is possible to note that the high-frequency switching control signal arises when the sliding variable becomes zero, as seen in Fig. 5.6b. Due to the observers, the closed-loop control system is free of chattering.

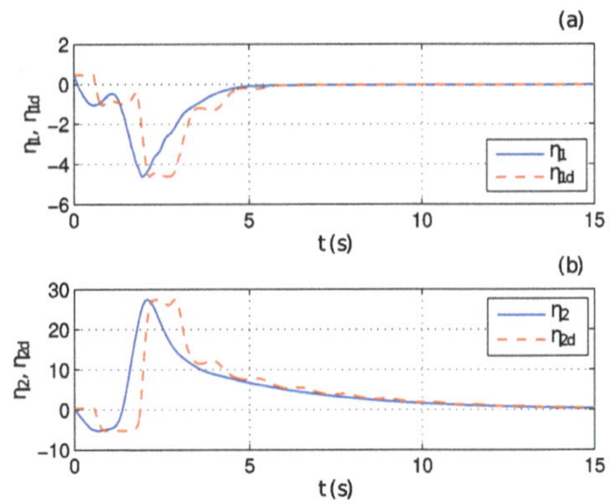

Fig. 5.5 Actual states η_1 (**a**) and η_2 (**b**), and delayed states η_{d1} (**a**) and η_{d2} (**b**)

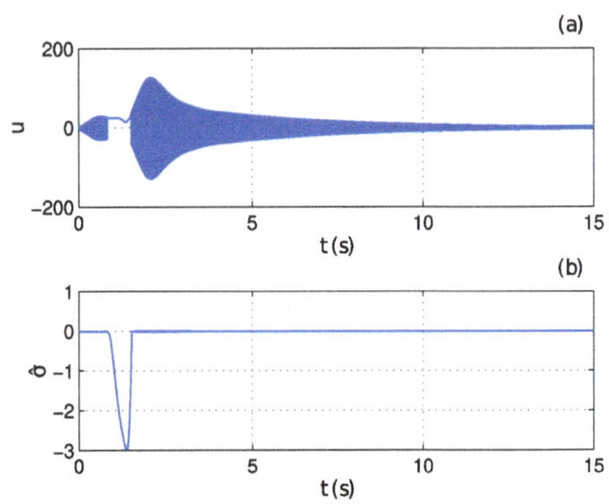

Fig. 5.6 Control signal (**a**) and sliding variable (**b**)

5.7 Application to Missile Guidance

In this section, we discuss the application of our technique to a missile guidance problem with output delay. A schematic diagram of the missile-target engagement is depicted in Fig. 5.7. The formulation follows the one presented in [19], where a planar linearized model of the engagement is considered. Alternative approaches and guidance laws can be found in the literature, such as [33] where sliding-mode controller and observer are adopted. It is assumed that the missile has already reached a near-collision course with a target, such that the line-of-sight (LOS) rate $\dot{\lambda}(t)$ is small. Also, the missile velocity ($v_M(t)$) and target velocity ($v_T(t)$) are assumed constant, in the horizontal direction and $v_M(t) > v_T(t)$. These conditions along with $\dot{\lambda}(t) = 0$ guarantee the target interception.

The closing velocity is thus constant, $V_{cl} = v_M(t) - v_T(t) = -\dot{r}(t)$, such that it is possible to write

$$r(t) = r(0) - V_{cl}t\,, \tag{5.72}$$

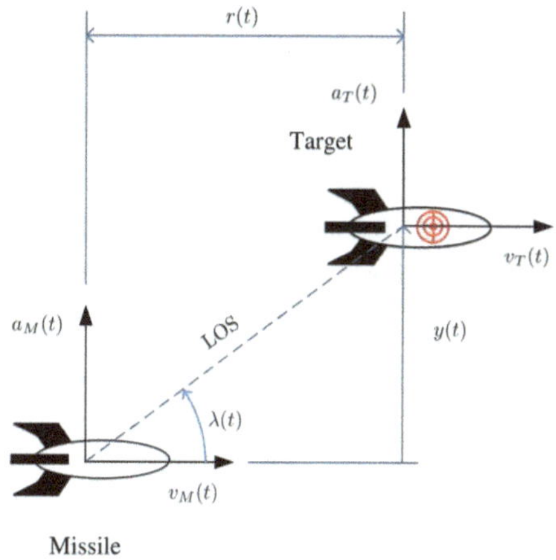

Fig. 5.7 Missile-target engagement diagram

where $r(0)$ is the horizontal displacement when terminal guidance begins [19]. Note that with constant velocity, engagement ends at the final time t_f given by $t_f = r(0)/V_{cl}$. Gravitational and aerodynamic effects are neglected in this model, such that the relative kinematics in vertical direction are solely determined by maneuvering accelerations $a_M(t)$ and $a_T(t)$, normal to $v_M(t)$ and $v_T(t)$, respectively. Thus,

$$\ddot{y}(t) = a_T(t) - a_M(t)\,. \tag{5.73}$$

The engagement model is then summarized as follows:

$$\begin{bmatrix}\dot{y}(t)\\ \ddot{y}(t)\end{bmatrix} = \begin{bmatrix}0 & 1\\ 0 & 0\end{bmatrix}\begin{bmatrix}y(t)\\ \dot{y}(t)\end{bmatrix} + \begin{bmatrix}0\\ -1\end{bmatrix}u(t) + \begin{bmatrix}0\\ 1\end{bmatrix}w(t), \tag{5.74}$$

where the state is given by vertical displacement and velocity, $x(t) = [y(t)\ \ \dot{y}(t)]^T$, the input signal is the missile acceleration $u(t) = a_M(t)$, and target acceleration is treated as a disturbance $w(t) = a_T(t)$.

The output signal presents a known constant pure delay d_x due to the measurement system:

$$y_d(t) = y(t - d_x)\,. \tag{5.75}$$

The control objective is to drive $y(t)$ and $\dot{\lambda}(t)$ to zero. In [19], it is shown that

$$\dot{\lambda}(t) = \frac{y(t)}{V_{cl}t_{go}^2(t)} + \frac{\dot{y}(t)}{V_{cl}t_{go}(t)}, \tag{5.76}$$

where $t_{go}(t) = t_f - t$ denotes the remaining time until engagement ends. Equation (5.76) configures a sliding surface in the form of (5.45)–(5.46) with

$\sigma = \dot{\lambda}(t) = 0$. This is implemented using observed estimates $\hat{y}(t)$ and $\dot{\hat{y}}(t)$, as follows:

$$\hat{\sigma}(t) = b_0 \hat{y}(t) + b_1 \dot{\hat{y}}(t) \,, \tag{5.77}$$

where $b_0 > 0$ and $b_1 > 0$ are constants chosen by design.

Note that when $\hat{\sigma}(t) = 0$, we have

$$\dot{\hat{y}} = -\frac{b_0}{b_1} \hat{y} \,. \tag{5.78}$$

This shows that the observer state variable $\hat{y}$ converges to zero. Recalling Lemma 5.2, we conclude that y also converges to zero.

This assures that the LOS rate is driven to zero when the sliding surface is reached. The control law is chosen as

$$u = -\text{sgn}(k_p)\varrho(\hat{\eta}, \hat{x}, t)\text{sgn}(\hat{\sigma}(t)) \,. \tag{5.79}$$

In this case, we have the following modulation function:

$$\varrho(\hat{\eta}, \hat{x}, t) = \left[\left| K_m \hat{x} \right| + k_x \|\hat{x}\| + \varphi_w(t) \right] / \underline{k}_p + \delta \,. \tag{5.80}$$

The disturbance $w(t)$ is assumed known. Notice that even though it is not possible to measure $w(t)$ directly, several estimation techniques that could provide a solution to this problem are reported in the literature, as found in [5]. We consider a rapidly maneuvering target described as in [2] with acceleration:

$$a_T(t) = \begin{cases} -80 \,, & \text{if} \quad 0 \le t \le 2 \,, \\ 0 \,, & \text{if} \quad 2 < t \le 4 \,, \\ 80 \,, & \text{if} \quad t > 4 \,, \end{cases} \tag{5.81}$$

in m/s^2, which is depicted in Fig. 5.8. A similar discussion is also seen in [31].

We consider that $r(0) = 2000$ m with a closing velocity of $V_{cl} = 200$ m/s, such that the engagement ends in 10 s. Similarly to [19], we have $y(0) = 0$ m and $\dot{y}(0) = 100$ m/s. An output time delay of $d_x = 0.4$ s was considered, such that the observers can be designed as described in Sec. 5.6, with $\theta = 4$ and $K = [0.85 \; 0.24]^T$. The controller parameters are: $S = [1 \; 1]$, $K_m = [-2 \; -3]$, $\underline{k}_p = 0.9$, $k_x = 0.6$, $\varphi_w(t) = |w(t)|$, and $\delta = 0.1$.

Simulation results are shown in Fig. 5.9, where the output and delayed output illustrate the difficulties induced by time delay in engagement dynamics. Note that the observer error is also driven to zero. In Fig. 5.10, it is possible to note that the sliding surface is reached. Also, $\hat{\sigma}(t)$ and the LOS rate $\dot{\lambda}(t)$ are driven to zero.

The control signal seen in Fig. 5.10a has high-frequency switching due to the ideal sliding mode allowed by the cascaded observers, which compensate for

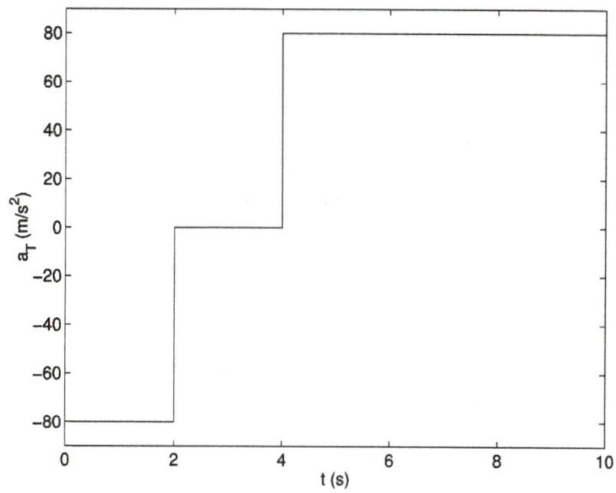

Fig. 5.8 Target acceleration

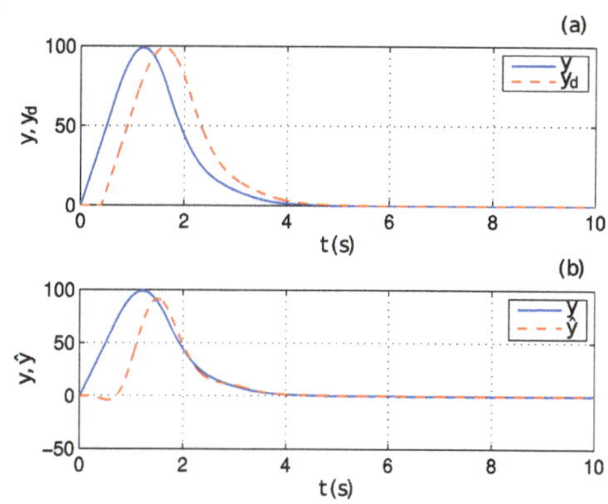

Fig. 5.9 Output signal $y(t)$ and delayed output measurement $y_d(t)$ (**a**), output signal $y(t)$ and estimated output signal $\hat{y}(t)$ (**b**)

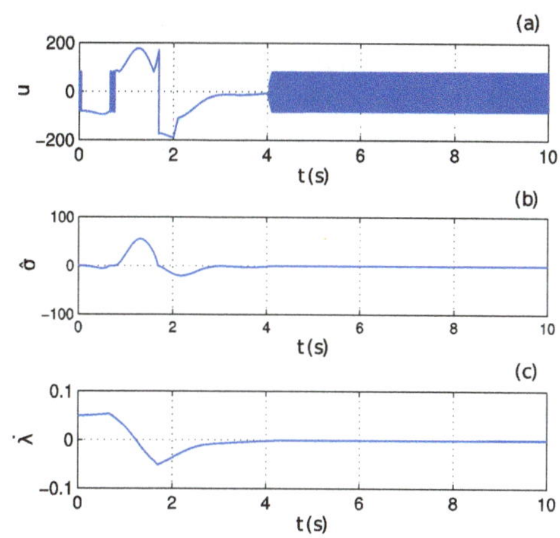

Fig. 5.10 Plots of acceleration $u(t)$ (**a**), sliding variable $\hat{\sigma}$ (**b**), and LOS rate $\dot{\lambda}(t)$ (**c**)

the effects of measurement delays. Thus, the chattering that would be caused by measurement delays, such as observed in Fig. 5.2b, is avoided.

5.8 Notes and References

The study of systems with time-delayed signals is of continuous interest of a wide variety of fields, such as biology, economics, mechanics, physics, and engineering, which include communication and information technologies, networked controlled systems, teleoperation, robotics, and pipelines, to name a few [29]. The presence of delays in control systems often leads to instability and poor performance [16, 32].

Sliding-mode control (SMC) is an attractive methodology for nonlinear systems, being robust to parameter uncertainties and disturbances [36]. On the other hand, actuator and sensor delays are among the most common dynamic phenomena that arises in control engineering practice [21]. In SMC, time delays deteriorate the control performance, since they cause chattering and may even destabilize the system. Despite of this, sliding-mode controllers for systems with state delays were proposed by Li and DeCarlo [22], Orlov et al. [27], and Gouaisbaut et al. [17] assuming full-state feedback. The use of state observers is an alternative for output-feedback stabilization of systems with state delay, as developed in [26, 38, 40]. However, such observers may not be applied to a wide class of systems. Adaptive stabilizers based on compensators (e.g., [4]) have well-known noise sensitivity.

Observers can be applied in SMC to avoid chattering caused by small time lags due to unmodeled dynamics in the measurement system [36, Sec. 8.3]. However, delayed output signals may impair the convergence of the estimated state to the true state, and, consequently, the closed-loop control system may become unstable.

Surprisingly, few results are available for SMC of systems with input or output delayed signals. Basin et al. [3] and Feng et al. [11] proposed SMC strategies based on state feedback for systems with delayed input signals as well as input signals free of delay, which facilitate the control. Liu et al. [23] applied Padé approximations to transform the SMC of systems with delayed output signal into the problem of controlling nonminimum phase systems. However, it is known that such approximations may be unrealistic for long delays. Furthermore, only local stability could be guaranteed for known parameters and known time delay.

Other recent works show interest in this topic. For instance, [28, 30, 39], and [12] deal with delay systems, however restricted to state delays only. Some approaches deal with both state and output delays, but may not be applicable when only output feedback is available or time-varying delays are present, such as [35, 37], and [34].

This chapter proposes an output-feedback SMC for nonlinear systems with multiple time delays, in the system state and output signal. The state delay is assumed time varying, and there is no limitation on the rate of its change. The output delay can be arbitrary provided that it is constant.

In addition, an innovative approach for the state estimation in systems with delayed output signals is the use of cascade observers [1, 15], which had not been explored in SMC.

In this chapter, the proposed controller is based on a cascade of high-gain observers [1] to compensate the delay effects in the system output signal and an asymptotic observer for the state of the zero dynamics. The combination of these estimators of two kinds allows the inclusion of state delays in the estimation problem not considered in [1]. Observers were previously adopted in [6] for the norm of delayed states and for systems with delayed output signal in [7]. Here, these techniques are applied to systems with delayed output signal, state delays, and zero dynamics.

Global asymptotic stability of the closed-loop system is guaranteed. It is important to stress that the scheme allows ideal sliding mode, even in the presence of delays. Indeed, the estimated sliding variable provided by the cascade observers becomes null after some finite time, therefore avoiding undesirable chattering phenomena.

The results reported here are significant when only the delayed output signal is available for feedback. As far as we know, proposing control schemes with global stability properties is a challenging task in the context of SMC for nonlinear systems with multiple time delays, without using any kind of approximation.

We considered an academic example of a nonlinear system with state and output delays followed by an application of practical interest, concerning the problem of missile guidance in the presence of seeker delays. In [10], this is motivated by the attempt to lower the costs of tactical missiles by employing low-cost optical seekers, which can present substantial delays in measurement of the missile-target line of sight. As discussed in [19], large seeker delays require compensation in the guidance law; otherwise, performance degradation or even instability can occur, ultimately resulting in target miss. Sliding-mode control is also applied to missile guidance in [31]; however, this technique does not deal with time delays.

References

1. Ahmed-Ali T, Cherrier E, Lamnabhi-Lagarrigue F (2012) Cascade high gain predictors for a class of nonlinear systems. IEEE Trans Autom Control 57(1):224–229
2. Babu KR, Sarma IG, Swamy KN (1994) Switched bias proportional navigation for homing guidance against highly maneuvering targets. J Guid Control Dyn 17(6):1357–1363
3. Basin M, Fridman L, Rodríguez-González J, Acosta P (2003) Optimal and robust sliding mode control for linear systems with multiple time delays in control input. Asian J Contr 5(4):557–567
4. Bobtsov AA, Pyrkin AA, Kolyubin SA (2014) Simple output feedback adaptive control based on passification principle. Int J Adaptive Contr Signal Process 28(7–8):620–632
5. Cloutier JR, Evers JH, Feeley JJ (1989) Assessment of air-to-air missile guidance and control technology. IEEE Control Syst Mag 9(6):27–34
6. Coutinho CL, Oliveira TR, Cunha JPVS (2013) Output-feedback sliding-mode control of multivariable systems with uncertain time-varying state delays and unmatched non-linearities. IET Control Theory Appl 7(12):1616–1623

7. Coutinho CL, Oliveira TR, Cunha JPVS (2014) Output-feedback sliding mode control via cascade observers for global stabilisation of a class of nonlinear systems with output delay. Int J Control 87(11):2327–2337
8. Cunha JPVS, Costa RR, Hsu L (2008) Design of first-order approximation filters for sliding-mode control of uncertain systems. IEEE Trans Ind Electron 55(11):4037–4046
9. Cunha JPVS, Costa RR, Lizarralde F, Hsu L (2009) Peaking free variable structure control of uncertain linear systems based on a high-gain observer. Automatica 45(5):1156–1164
10. Dhananjay N, Lum KY, Xu JX (2013) Proportional navigation with delayed line-of-sight rate. IEEE Trans Control Syst Technol 21(1):247–253
11. Feng Y, Yu X, Zheng X (2006) Second-order terminal sliding mode control of input-delay systems. Asian J Contr 8(1):12–20
12. Figueredo LFC, Ishihara JY, Borges GA, Bauchspiess A (2013) Delay-dependent robust stability analysis for time-delay T-S fuzzy systems with nonlinear local models. J Control Autom Electr Syst 24(1):11–21
13. Filippov AF (1964) Differential equations with discontinuous right-hand side. Amer Math Soc Transl 42(2):199–231
14. Fridman E (2001) New Lyapunov-Krasovskii functionals for stability of linear retarded and neutral type systems. Syst Contr Lett 43(4):309–319
15. Germani A, Manes C, Pepe P (2002) A new approach to state observation of nonlinear systems with delayed output. IEEE Trans Aut Contr 47(1):96–101
16. Gomes da Silva Jr JM, Bender FA, Tarbouriech S, Biannic JM (2009) Dynamic anti-windup synthesis for state delayed systems: an LMI approach. In: Proceedings of the 48th IEEE Conference Decision and Control held jointly with the 28th Chinese Control Conference, Shanghai, pp 6904–6909
17. Gouaisbaut F, Blanco Y, Richard JP (2004) Robust sliding mode control of non-linear systems with delay: a design via polytopic formulation. Int J Contr 77(2):206–215
18. Han X, Fridman E, Spurgeon SK (2010) Sliding-mode control of uncertain systems in the presence of unmatched disturbances with applications. Int J Contr 83(12):2413–2426
19. Holloway JC, Krstic M (2015) A predictor observer for seeker delay in the missile homing loop. In: 12th IFAC workshop on Time Delay Systems, Ann Arbor, MI, vol 48, pp 416–421
20. Khalil HK (2002) Nonlinear systems, 3rd edn. Prentice Hall, Upper Saddle River
21. Krstic M (2009) Delay compensation for nonlinear, adaptive, and PDE systems. Birkhäuser, Basel
22. Li X, DeCarlo RA (2003) Robust sliding mode control of uncertain time delay systems. Int J Control 76(13):1296–1305
23. Liu G, Zinober A, Shtessel YB (2009) Second-order SM approach to SISO time-delay system output tracking. IEEE Trans Ind Electron 56(9):3638–3645
24. Michiels W, Niculescu SI (2007) Stability and stabilization of time-delay systems: an eigenvalue-based approach. SIAM, Philadelphia
25. Nam PT (2009) Exponential stability criterion for time-delay systems with nonlinear uncertainties. Appl Math Comput 214(2):374–380
26. Niu Y, Lam J, Wang X, Ho DWC (2004) Observer-based sliding mode control for nonlinear state-delayed systems. Int J Syst Sci 35(2):139–150
27. Orlov Y, Perruquetti W, Richard JP (2003) Sliding mode control synthesis of uncertain time-delay systems. Asian J Control 5(4):568–577
28. Ramakrishnan K, Ray G (2015) Improved results on delay-dependent stability of LFC systems with multiple time-delays. J Control Autom Electr Syst 26(3):235–240
29. Richard JP (2003) Time-delay systems: an overview of some recent advances and open problems. Automatica 39(10):1667–1694
30. Shi P, Zhang Y, Chadli M, Agarwal RK (2016) Mixed H-infinity and passive filtering for discrete fuzzy neural networks with stochastic jumps and time delays. IEEE Trans Neural Networks Learn Syst 27(4):903–909
31. Shtessel YB, Shkolnikov IA, Levant A (2007) Smooth second-order sliding modes: missile guidance application. Automatica 43(8):1470–1476

32. Si-Ammour A, Djennoune S, Bettayeb M (2009) A sliding mode control for linear fractional systems with input and state delays. Commun Nonlin Sci Numer Simul 14(5):2310–2318
33. Song J, Song S, Guo Y, Zhou H (2015) Nonlinear disturbance observer-based fast terminal sliding mode guidance law with impact angle constraints. Int J Innov Comput Inf Control 11(3):787–802
34. Su X, Shi P, Wu L, Song YD (2013) A novel control design on discrete-time Takagi-Sugeno fuzzy systems with time-varying delays. IEEE Trans Fuzzy Syst 21(4):655–671
35. Su X, Shi P, Wu L, Basin MV (2014) Reliable filtering with strict dissipativity for T-S fuzzy time-delay systems. IEEE Trans Cybern 44(12):2470–2483
36. Utkin V, Guldner J, Shi J (1999) Sliding mode control in electromechanical systems. Taylor & Francis Ltd., London
37. Wu L, Su X, Shi P (2012) Sliding mode control with bounded $\mathcal{L}_2$ gain performance of Markovian jump singular time-delay systems. Automatica 48(8):1929–1933
38. Yan XG, Spurgeon SK, Edwards C (2010) Sliding mode control for time-varying delayed systems based on a reduced-order observer. Automatica 46(8):1354–1362
39. Yan XG, Spurgeon SK, Edwards C (2014) Memoryless static output feedback sliding mode control for nonlinear systems with delayed disturbances. IEEE Trans Automatic Control 59(7):1906–1912
40. Yan XG, Spurgeon SK, Orlov Y (2014) Output feedback control synthesis for non-linear time-delay systems using a sliding-mode observer. IMA J Math Contr Inf 31(4):501–518

FSC
www.fsc.org
MIX
Papier | Fördert gute Waldnutzung
FSC® C083411

Zeitfracht Medien GmbH
Ferdinand-Jühlke-Straße 7
99095 Erfurt, Deutschland
produktsicherheit@kolibri360.de